CW01082883

BLUEPRIN' TIME MACHINE

ROBERT STARK

outskirts press

Blueprint for a Time Machine

v1.0

Outskirts Press, Inc.
http://www.outskirtspress.com

Paperback ISBN: 978-1-9772-3702-6
Hardback ISBN: 978-1-9772-3827-6

PRINTED IN THE UNITED STATES OF AMERICA

Table of Contents

PART ONE

"I'd like to go there again, to feel the water bubbling over my head, and wash over my bare shoulders. To feel it swirl around my legs and run between my toes, washing away the heat of the day. To smell the freshness and taste the moment forever. Yes, even to be whistled home by my father… Can I go back to that moment?"

Chapter 1

Jason Strider sat at the wooden bar and slowly returned from his memories. He polished the tiger stripes in the old quarter-cut oak bar with the sleeve of his worn denim shirt. He looked across to the mirror and lifted his glass of draft beer in a mock toast to his image.

He looked around the familiar room and felt comfort in all that surrounded him. The pool table with worn green felt tattered from years of heavy use. The cheap Naugahyde upholstery on the booths and bar stools. The familiar faces of some of the people from the place he worked. The women at the booth across from the pool table, seriously engrossed in their conversation. He was twenty-seven and could feel the years creeping up on him. He looked into the mirror on the other side of the bar and saw the lines around his eyes crinkle into a smile. They sparkled with the remembrance of youth. Not bad, he thought to himself. One kid at home, a good job at the factory, a good car, a pretty wife...

He watched his smile fade as reality was dragging him down

The dark stranger sitting at the bar next to Jason patiently waited. He had sat down and spoke but received no response. He at first thought that Jason was ignoring him, then realized there was nobody home. The body was there, the eyes were open, but there was nobody home. He smiled inwardly to himself. Most would think this fine-looking fellow with a strong jaw and fine features was just some drunk but LeRoi knew different. LeRoi ordered an Ouzo and waited.

Jason sat at the wooden bar and slowly returned from his memories. He downed the rest of his beer and asked for another, trying as always to keep himself from falling into a bad mood. He hated to drink when he was depressed so he forced himself to look for another good memory.

"Talk yourself up." He thought. "Look for the positive."

He looked at his image again and what he saw of his physical being pleased him. But he also felt the reality of his homelife. He could see the pectoral chest muscles beneath his shirt. He unconsciously flexed and watched the wonder of his body in the mirror. He knew that although his physical being was above average, his marriage was falling apart.

His eyes looked in the mirror at the guy on the stool next to him. Their eyes met. He half blushed at being caught admiring himself. He looked away from the mirror and down at his beer.

The stranger's eyes were dark. His hair was jet black, his face kind, but expressionless. Jason glanced at the stranger's hands and noticed a stub where his ring finger used to be. He quickly looked back to his own hands and said a silent prayer of thanks for all his fingers were still in place.

The standing joke at the copper factory where he worked was that a prerequisite for advancing to foreman was losing a finger. Logically he knew that wasn't the case. But realistically, he knew that most of the bosses were short a finger or two. He thought back to the copper factory.

Manufacturing copper wire required pulling a wire through a hole in a die. The process was always the same. Take a spool of wire, put a point on the loose end, stick the pointed wire through a die and wrap it around a power-driven pulley called a pull block. Pull another point on the wire, stick it through a smaller die, wrap it around another pull block. On it went through ten or eleven dies and then the wire was fastened to an empty spool. Each pass of the wire through a die made the length longer, and the diameter smaller. Fine wire, smaller than a human hair, was drawn through holes in diamond dies, but the big stuff was drawn through holes in Tungsten dies. The bigger the wire, the more powerful the pull-blocks. Each pass of the heavy wire through a die provided a pinch point for careless fingers.

As he sat at the bar, he wiggled all his finders, thankful again that he had not yet had that mental lapse or that moment of carelessness which would provide another meal for an ever-hungry wire drawing machine.

The dark-haired stranger broke the silence. "Still got'm all I see."

Jason was again embarrassed at being so obvious.

"Yeah. I'm not ready for a foremen's job yet."

The stranger's dark eyes looked confused, and Jason could see he wasn't from the copper factory.

"Just a private joke, Jason said, he then swiveled around on his bar stool and pretended to watch the pool game.

The stranger made him feel a little uncomfortable, but he also felt a warmth he seldom experienced, Jason downed his seventh beer, and ordered another draft. He could feel himself slipping into another day-dream. The alcohol seemed to help him into that most enjoyable part of his past- his childhood. He pretended to watch the pool game. His eyes partially glazed over, and a slight smile crossed his lips as he re-called a piece of a different era.

The grade school was an old red brick building with ramps that led down to the coal-fired furnace. There were classrooms down in the lower area for the first two grades of elementary children. Jason had spent his first year of school there and was now on summer break. Next to the school was an old red barracks that had been converted into a branch Library. Jason was seven years old. The school playground with slides, teeter-totters, swings and merry-go-round was an adventureland for the young Jason Strider. The entire playground was drained of its rainwater via one long shallow storm drain which led to a storm sewer. It didn't take much imagination to figure out that if the storm sewer grate were plugged, a rainwater, stream filled, mini lake would appear.

This Saturday was hot and the black storm clouds had been gathering for hours. Young Jason gathered some old paper sacks and leaves. He tore an old grocery sack into medium sized pieces. Each piece of the sack was placed in just the right spot to cover the voids in the metal grate. Next, sand was added to hold the first layer of paper in place, and then small rocks to solidify and hold the sand. The rain came in big splattering drops. Jason ran to his favorite spot beneath the overhang of the old barracks. The hole in the rain gutter above his head provided him with his own waterfall and soon he couldn't keep from it. The water gushed over him in a torrent, bubbling over his fine brown hair, over his shoulders, almost washing his suspenders off his bare back. It cooled and refreshed him. The rain began to let up and Jason headed for the culvert to examine his handiwork.

Quickly he rolled up his pants, not so he could save them from the water, but because he wanted to better feel the muddy culvert water on his slim legs. Into the fast-running stream, he waded; dancing and kicking water in every direction. Upstream he ran, high stepping and splashing all the way to the mouth of his private river. The rain was over now but not the fun. Gathering a block of wood, he pretended to launch a boat and instantly, he became a sailor. With an old tree limb for a sword, he slashed and sailed his way into deep water. He felt the pull of the ocean and all the creatures it held.

He heard a shrill whistle and in his mind's eye he could see his father standing on the back porch, two fingers in the corners of his mouth, sounding the whistle that was calling him home for dinner. Never enough time, he thought as he waded back to his mini dam. His small toes felt for the rocks and sand and paper that were holding back the river. Ever so gently, he pulled away a few stones and sand. With his big toe, he punctured the soft paper, and gleefully felt the vortex begin to swirl around his legs. Bits of sticks bumped into his legs like little fishes coming in for a nibble. Sand and mud caressed his feet giving him a feeling he would never forget.

"Crack!" the sound of the cue ball smashing into the triangle of object balls brought him back to the present. At first, he was disorientated and confused. He looked around the bar room and realized where he was. God, what a feeling, he thought. He looked down at his feet, somehow expecting to see tiny toes dripping with muddy water. Jason felt his shirt and pants expecting to feel rain-soaked clothes, but his hands came away dry.

"Still dry I see."

His mind started to do a quick trace back, but it found nothing.

"I must of be talking to myself," he thought. This guy heard me talking and reasoned that I was dreaming about the rain.

Now Jason was really confused. He looked around again to see if anyone else in the bar was watching. No one else seemed to have noticed. People playing pool, talking and drinking. He looked back to the dark eyes of Mr. Four Fingers.

"Le Roi Verita," the man said as he held out his hand. "You're probably wondering how I could know where you were?"

Jason took his hand and felt the strength and warmth of...well, it felt like friendship. How could that be? You can't feel friendship ...or can you? Sure, after enough beer, you can probably feel anything, he thought.

"Jason Strider," he said, "what ya drinking?"

"Ouzo."

Jason turned to the bartender and said, "Give us two Ouzos."

He didn't know what Ouzo was, but if it could give you the power to read minds, he wanted to try some. The bartender poured the shots and both men picked up the colorless liquid, made a silent toast with their eyes, and both downed the contents. It was strange tasting, smooth and sweet. But everything was beginning to seem strange to Jason.

"OK, I'll bite, how'd you read my mind?"

"I didn't."

Still confused but also feeling so good it didn't matter Jason asked, "You mean we somehow traveled back to my old grade school together?"

"I couldn't feel the same things you did, but I could watch you. It looked like great fun."

Jason motioned to the bartender and ordered another two drinks and took a small sip. He was trying to get a handle on what this LeRoi guy was talking about. The only thing that came to mind was the names of movies he'd seen on time travel. Orson Wells, "Time Machine" and "Journey's Past". What great names, but all that was fiction. This guy was talking like you could really do it.

"Are you talking about really traveling back in time? I mean really traveling back?"

"Yes," replied LeRoi with a nod.

"OK, let's do it again...I mean...if you don't mind. If it can really be done, let's do it again".

LeRoi was trying to stay calm and act like it was as common and ordinary as breathing but he could hardly contain his excitement.

The questions were flooding into LeRoi's mind. Was this the student he'd been looking for? Was this the one that would help him along the trail of his seemingly endless journey? How could this uneducated bar patron have matched his thinking so perfectly?

LeRoi said in what he hoped was a nonchalant voice. "Yes, it's possible, but it takes some practice to be able to control it. Would you like to learn?"

Jason was hesitant, should he tell him he was a high school dropout? Should he tell him he wasn't capable of doing much more than simple algebra? That his English and spelling was mostly limited to four letter words and lousy grammar.

"What kind of learning are you talking about? How much will you charge me? Where are the classes?"

"It won't be like any class you've ever had," LeRoi replied.

He knew he had to play it exactly right. A wrong move and he could lose him. He saw the doubt creeping into Jason's Eyes.

"It will cost you a few hours on Friday nights, and we can learn right here or over in one of those booths. The first lesson can be right now if you like. We'll go as fast as you like and when you become bored, we'll stop."

"How long will it take?" Jason was interested, but this seemed too good to be true. Free lessons in time travel. Ha! Wait till he told the guys at work… or would he tell them?

"What am I thinking? This is impossible… But I did just go back." He thought for a while. "Oh, what the hell. Sure, I'm game. Let's give it a try."

LeRoi picked up his Ouzo and motioned towards a corner booth. They slid into the comfort of a dark corner in of the back of the bar.

"Well, let's start at the beginning."

"The beginning of what?" Jason asked.

"The beginning of time of course. In order for you to understand time we'll have to start where it all began… at least in the beginning of earth time… Real Time."

"You mean we'll have to go all the way to the beginning of the earth? No one knows what that was like."

The argumentative part of Jason's brain kicked into gear despite the alcohol. He looked across the table and saw a glance from LeRoi that took him back to his second-grade teacher. The 'shut up while I'm talking,' look.

LeRoi softened his glare and silently chided himself for being impatient. This is going to be a struggle, he thought. So much to teach, and so much to learn. He thought of the old saying, A journey of a thousand miles...

LeRoi started out slowly: "It came together two billion years ago. Just a mass of elements, a tremendous rush of gravitation and friction and adhesion. It was a great spectacle of birth. Mother Earth as a baby was quite a sight to see. Fifth in size among her sister planets, weighing in at sixty-six sextillion tons with a waistline of 24, 901 miles. The home of man, and distinct from heaven, hell and all other celestial bodies."

His definition had come almost word for word from the old Webster 's Dictionary. He looked at Jason and saw the interest in his eyes. How often he had seen it work? Throw a few hard facts at the uneducated, and they begin to drop their guard. How many times had he tried to break down the mental walls that he always encountered? How many times had he failed? Kicked out of two universities. Fired from three jobs, and now, sitting here talking to...a what... well, he didn't know. But he decided to drive forward and attempt to make his first point.

"The fiery mass began to cool, and the phenomena known as Bowen's Reaction Series began."

"What is Bowen 's Reaction Series?"

"It just means that as the earth cooled, some elements like bedrock solidified first and other elements that have a lower melting point solidified later. Separation of elements created different deposits in different places. It's a simple explanation of how things probably were in the beginning."

"If you know that, then you could figure out where the gold and diamonds and silver are located! We could go out and pick up as much as we wanted!" Jason was beginning to like this.

LeRoi ignored the remark and continued. "The point I'm trying to make is that the earth began and was born just as you were born. Each year is an earth birthday just as each year you have a birthday. **'Real Time' began and is going in a straight line. Each year is a year unto itself, one revolution around the sun. Each minute is a minute unto itself. It happens and it cannot happen again. It cannot be done over!"**

Jason got that confused look on his face again. "First you tell me you can teach me to travel in time and then you tell me that one minute is a minute unto itself, and once it's over it's over and it cannot happen again. That kind of makes time travel into the past impossible, doesn't it?"

"One thing at a time," LeRoi thought.

"I'm talking just about Real Time, and it is constant. Once it occurs, once that minute passes, it cannot happen again." He saw the disappointment in Jason's eyes, but he knew he had to go slow. There were only a small number of points to be made, but in the past, he'd gone too fast, encountered too much resistance, lost too much credibility. "Yes, he thought, one lesson at a time."

"Real Time is measured on that clock over there. It's geared to revolutions, or cycles of the earth. The earth was born, she lives out each year and after a certain number of cycles she will die, just as we must die."

"A bit melodramatic," he thought, but he was on a roll. LeRoi glanced across the table to read Jason's eyes and realized he was talking to himself.

The workday and the beer and the Ouzo had been too much. His new student was slumped against the wall, chin on his chest, breathing the slow easy rhythm of sleep. LeRoi slipped out of the booth and headed for the back door. The night was cool and black. He glanced around to see if anyone was watching and quickly slipped into the heavy brush.

The bartender, Frank, glanced out the back window and saw a dark form disappear into the brush. He shrugged his shoulders and went back to rinsing glasses. Some guys would always prefer bushes to restrooms, he reasoned.

Quietly, LeRoi felt his way along the game trail leading down the steep bank to the river. His only friend, the river, deep and strong. The sleek Sawyer canoe waited. He reached for the gunnels. The brushed aluminum was cold to his touch. One foot in the middle and an easy push with the other and he slipped into the slow current. He eased onto the boat cushion in the middle of the canoe and picked up the lightweight, laminated, butterfly paddle. Instead of

paddling he merely dropped his paddle aft and used it to rudder himself to midstream.

The feeling was always the same, it was like stealing a free ride. Two miles to home and all he had to do was steer. The still water along the edge of the river had allowed him minimum effort in getting upstream to the bar, and now the strong middle current would take him home.

His mind traveled back to the bar and he could almost see his new friend asleep in the corner.

What was he like? What could he teach him? What would the change bring? Would it help him, hurt him, kill him? Would he ever Even see him again? Why had they met?

He knew some had gone on time journeys and never returned, or at least he suspected as much. Not Entirely his fault he reasoned. After all, it was Jason who had been doing the traveling, and it wasn't the first time. This guy was an expert traveler, even if he didn't know what he was doing. Where might they be able to go with a little bit of knowledge?

"Science and adventure were not without their risks," he thought.

The bartender was wiping off empty tables and placing the chairs on top. "Ashtrays, motherfuckers!" he called out. He turned up the lights and continued to put the bar in order. The bright lights brought Jason out of his daze. He looked around and realized it was just him, a few other drunks and the bartender. He got up and headed for the door. Three AM and just enough sleep to make him feel like shit. He got into the car and headed for home.

The house was dark. "Hope she's asleep," he thought. Up the stairs and quick look in on his son. Off with his clothes and into bed.

She was on the far side of the kind-sized monstrosity. "She might as well be in another county," he thought. No food, no sex, and the house was a mess.

"Drunken jerk," She thought. "Why couldn't he be more like my father?"

His last thought before sleep surrounded him was of a four fingered stranger named LeRoi. Was he real or just an apparition?

Saturday dawned much too early for Jason. His son was bouncing on the bed. His wife was already up and watching TV. Through the fog from within his aching head he looked at his son. Seven years old and growing up too fast.

"Let's go out and play catch Dad."

"Okay, just get me a cup of coffee and let me wake up some."

Off Brian ran with all the bounce and recklessness of youth. Jason's only joy in life was his son. "What a world," he thought. Work and drink and play a little, and then start the cycle all over again. Stack weeks on top of weeks and the year's race by, but for what? What's the use? His head began to pound as his son appeared in the doorway, carefully carrying the hot fluid that would partially renew him.

"Come on Dad."

Jason rolled out of bed, practically tripping on his blue jeans and shirt. He picked them up and caught a whiff of the bar. Stale smoke and sweat. His mind flashed back to LeRoi, but he pushed the thoughts away and headed for the shower. The hot water revived him.

A quick game of catch, and they headed for the park. Jason marveled at how little effort his son demanded. One mile of walking for Jason was about two miles of running for his young son Brian.

"What kind of tree is this? Did you see that yellow bird? Why is it yellow? Look at the butterfly. Why is it orange and black? Where do they go in the winter? Could you swim across the river Dad? Where is all the water going? Why are bees so mean? Why is that dog poop on the trail? Girls are dumb aren't they, Dad? Why is the skin broken on that tree?

Jason walked over and looked at the tree. Seldom did Brian wait for an answer. Not that Jason had any. He only knew that if life was a question, this boy was an answer.

The tree was an old white oak. It was four feet around and tall enough to pain your neck. Jason lay down on his back and settled into a soft patch of new grass as Brian gathered stones for skipping. He let the spring sun saturate his being. He looked up again at the old tree. Its skin was indeed broken. Evidently a recent lightning bolt had blown a long vertical chunk of ragged bark from the top to the ground. He drifted...drifted back till he was the same are as his son. Perfect communication, he thought, was no words at all. Just a feeling and today was the only day in the world. A minute unto itself.

Chapter 2

Monday morning, the alarm shocked Jason into consciousness. Jason's Monday didn't seem so bad this time. Seven years at the copper mill, and today he was moving into a new department. He'd worked all but one, and today he was moving into the final unknown, the rolling mill.

He was both fearful and excited… as excited as he could be over work anyway. He looked forward to the new week.

Light blue denim shirt, old blue jeans, and tennis shoes. He quickly went out the door. Four and half miles in the old Pontiac and into the plant parking lot. Jason waved at the big black security guard as he passed the guardhouse, and into a familiar parking space. Up the cement steps and into the entrance with long green metal racks full of white computer punched timecards. He made a quick scan of the cards and pulled his from the many.

Jason listened to the time clock ding as he inserted and removed his card. Six thirty-three AM registered in blue ink on the "IN" column. He automatically slowed down once the timecard was punched. He inserted his card in the proper holding slot and began his long walk down the factory aisle to the locker room. The same metallic smell reminded him as always of the things he had learned. Factory smart was what he was. He remembered the first time he'd walked down this aisle. Always the tremendous glitter of bright shiny copper everywhere;

spools of copper wire. They were every size, five pound, ten, twenty, one thousand, five thousand pounds. Wire as small as a hair, up to rod as big as your thumb. Noise...noise everywhere, wire drawing machines, overhead cranes, acrid smoke from the pickling operations. He thought he'd never get used to the glitter, but he had. Now most of it just reminded him of work.

"Hey Patch, they're going to burn your ass today", someone yelled from the mass of equipment.

"What a handle," Jason thought. The first week he'd worked, he had worn an old pair of blue jeans he had patched himself. Both knees and the rear pocket had been reinforced with his own sewing. One of the red necks had begun calling him Patch. The more he'd complained, the more the new name had stuck.

First rule of the factory, he remembered, never let the assholes know what pisses you off or you will get a heap of it.

A forklift slid around the steel floored corner, carrying a six-thousand-pound coil of shiny, freshly annealed copper rod. Jason automatically stepped out of the way and let the truck pass. He continued down the aisle toward the locker room.

First fine wire, then intermediate wire, coarse wire, and then past the rolling mill. Jason walked past another time clock (this one for the more senior factory rats), and through the metal locker room door. The locker room was already steamy from the third shifters.

Naked men in the shower, laughter and small talk. The locker room smells that reminded him of high school football filled his senses. Sweat, mildew, soap and steam.

He inserted his key into one of the small lockers along the wall and extracted a fresh pair of work clothes. It took about two minutes to change from his blues to his works. Off with his tennis shoes and into his steel toed leather work boots. They felt cool and stiff from their two day's rest. He quickly laced them up. He left his pliers and holster in the locker. He'd be getting new tools today. New department, new job, new people to work with.

Each of the previous departments had been similar. Learn the jobs, learn the equipment, lose interest and become miserable. Come to work late, take a sick day every time he could get away with it, hate the boss, hate the company policy, the injustice, the noise, the dirt, the smells, then move to the next job. New interest, new goals...would this one be any different? I hope so, he thought, this is the only one left.

He headed for the Superintendent's office, a narrow staircase leading up to a glassed-in area. The office was cluttered. Two desks were piled with stacks of paper. The Super, Ken Morehead, was on the phone.

"Get a hold of Fred before he gets in the shower and hold him over two hours. I'm short a lineman and a sticker. I'll use Fred on the Hill and work short on the line 'til those drunken bastards get in here. I got one foreman to give breaks, and I got new meat to train today." He swiveled around and gave Jason a disgusted look. "Yeah, we'll make do. I want this fucking mill rolling by seven AM.

Jason glanced at the clock and noticed it was just starting time. "Six forty-eight, Real Time," he thought. Jason looked out the window that spanned the front of the office. The rolling mill was laid out below him. It was half as big as a football field and cluttered with giant mills of steel.

The over-head crane which ran the width of the plant on overhead rails was slowly moving away towards the back of the factory. He could just make out the crane operator through the blue haze.

Positioned in the cabin of the crane, the operator was driving towards the bar dock to pick up a load of copper. Bars on the back dock were stacked like cord wood. Six to a layer, eight layers to a stack. Each bar weighed two hundred and forty-five pounds, each stack, five ton.

The crane operator positioned the derrick above a stack of bars and the crane chaser slipped the great steel chains around the five-ton load. He motioned to the crane operator and the slack disappeared from the chain. The chaser walked around the stack and made sure all the chain was in the proper places. He gave the 'all clear' sign to the operator.

The load swung free of the ground and centered under the derrick. What happened next was a poetic dance between the operator and the load. With a mating of mind and machine, the stack floated up, over and away towards the waiting furnace operator. With hands and feet tweaking each control with just the right amount of finesse, the operator gently brought the stack of cold bars above the equipment, weaving and dancing the load through a maze of rolling mills, to a position adjacent to the fiery, doughnut-shaped, gas-fired furnace. He locked his controls and leaned back in his chair and waited for the chaser to catch up. To Jason, it was a feat of enormous skill, but to the crane operator and chaser it was just the beginning of another long day.

Jason looked back to the Super and watched him push the disconnect button on the telephone. Ken dialed a three-digit code and waited for the person on the other end to pick up the phone. "Let me talk to Al Bunhke...Al, this is Ken. Got a new man here name of Patch. How about showing him around? ...Good, see you in about five minutes."

Jason knew Al. He'd seen him many times in the shower. He was an old-time sticker who was about ready to retire, that is, if he didn't die first. He always appeared to be on his last leg. Ken Morehead turned his attention back to Jason.

"So, you're Patch."

"Jason Strider's the name."

Ken ignored the statement, as if Jason had never spoken.

"I talked to your old Super, Bob Sweet, in the wire mill this morning. What do you think he had to say about you?"

Jason cringed inwardly. "Probably wouldn't use him for a reference," he said, "or my wife either," he thought.

"He said you're a smartass and an asshole and a troublemaker. Said you couldn't be depended on to work more than three days in a row and you can't make it to work on time or sober."

Jason stood there and shifted his weight from one foot to the other. He thought about the new rule that forbid union and company men from swearing at each other. He smiled to himself knowing that in a one-on-one situation it could never and would never be enforced. He didn't really give a good fuck what this jerk-off superintendent thought about 'him. He'd gotten this job through the seniority bid system. If he could do the work, he could keep the job.

He looked at Ken and smiled. Ken sat there with the veins standing out on his temples. His once black hair had receded back to the rear of his head. A cigarette smoldered in an ash tray full of butts. Little beads of sweat were forming on his upper lip. His hands and wrists were scarred with burns. A scar ran through a sideburn and across one cheek. Black stubble was already showing through his morning shave and his face was lined with too many years of worry and pressure. His left hand, Jason noticed, was two fingers short of a full complement. "One for the foreman's job and one for the superintendent's job," he thought to himself. "I'll outlive this cocksucker."

Ken continued. "There are only two kinds of jobs in this copper factory. The rolling mill and everything fucking thing else. One guy fucks up and every one of us loses money. We work as a single unit back here. Our only objective is to get the highest number of bars rolled as possible."

"Safety was obviously not a priority," Jason thought.

"Nothing gets in the way of that goal. If you can't do the work, I won't need to run your young ass out of here, they will." He stabbed his thumb in the direction of the rolling mill. "You fuck up bad enough, they won't run you off, they'll burry you, or better yet, you'll burry yourself so deep in hot copper, won't nobody be able to save you."

Jason hadn't really needed to hear all this. He was already scared of the job. The rest of the plant thought the mill men were crazy. The first time he'd seen the mill people work he was awed by their fearlessness. They were handling cherry-red, hot copper rod with only a short pair of tongs. That was years ago and now here he was, about to step into the fire.

He saw old Al slowly making his way up the narrow staircase. He was relieved at having the opportunity to get away from this insulting company bastard.

Al Came through the door. "Al, this is Patch. Show him around the mill. Get him some tools and turn him over to Big John." He looked at Jason. "You got three days patch. Learn the job or get the fuck out!" Jason and Al headed out the door.

"Nice guy. Is he this personable to everyone or am I getting some sort of special treatment?"

Al Stopped and gave Jason a long look. He took a deep breath as if to start a long dialog but just shrugged his shoulders and slowly, placing one old gnarled aged hand on each rail, led the way down the steel staircase.

Each step was a careful concentrated effort. His frame reminded Jason of a loosely filled, thin lined, bag of bones. What little hair he had left was pure white. His shoulders were bony and sharp. His gigantic feet fit sloppily into brown leather shoes. His face sagged in every conceivable place as gravity gently pulled the baggy folds of skin down into rivulets of wrinkles.

They made their way to the store's window. The translucent, scratched plexiglass window was in the down position. Al raised the sliding window revealing the store's clerk at the counter. A newspaper lay on the counter. A cup of steaming hot coffee sat beside the paper. Not the machine kind the rest of the factory rats drank, but the fresh brewed kind from his personal coffee pot.

"Stub, this here's Patch. How's about a couple cups of coffee?"

Stub looked at Jason like he was the biggest intrusion to ever show up at his window. He grabbed a couple of paper cups from beneath the counter and filled them both.

Jason automatically looked at Stub's fingers. All ten he noticed. Maybe they called him Stub because of the short chunk of cigar in his mouth that was being chewed to a mushy pulp. Or maybe he had a short dick. He smiled inwardly at his thought but said nothing. "Don't be a smart-ass and don't get off on the wrong foot again. Don't fuck this last chance up," he reminded himself.

"Need a set of tongs for Patch. He'll be starting at the first station."

Stub left for a few moments returning with an armful of tongs. Each pair looked alike to Jason.

Each were two feet long, smooth handled and short jawed, 'black smith' type tongs for handling hot copper rod. Each of the jaws in the new tongs were shaped to accommodate specific sizes of copper. Jason picked up is coffee and took a precautionary sip. Al picked up each of the tongs in turn, held them to the light and worked them open and shut a couple of times.

"Got any of the old malleable iron type back there?"

Stub walked away grumbling but came back with three more pair. Al finally selected what seemed to be an appropriate pair and thanked Stub for the coffee. He handed the tongs to Jason and headed for the furnace at the rear of the mill.

The cold steel of the tongs felt good to Jason. He felt better now. Cold steel in one hand and hot coffee in the other, like a cowboy with a new gun he thought. They stopped at the huge doughnut-shaped furnace.

"Jake, this is Patch. I'm starting him out here with you."

Jason looked surprised. "I thought the super said I was starting on the Hill?"

"Ken gave you to me and I'm starting you here with Jake. You'll spend a week with him, and he'll show you how the mill operates."

"What about the three day's break-in time he said I had?"

"You have thirty days, just like the contract says."

Al walked over to a footlocker and pulled out a twelve-inch long, nine-sixteenth inch, oval-shaped piece of solid black copper. He reached over and took Jason's new tongs.

He held the tongs in one hand and threw the heavy piece of black rod into the air. Jake leaned back against a scrap barrel as if an announcement were about to be made.

Jason stood and stared into the old man's eyes. It was as if a transformation had suddenly taken place. Al's eyes took on a sparkle and a clarity. It was like he was a kid again, showing off an old favorite toy. Al reached for the falling chunk of the of metal with the new tongs. He caught the rod and swung down with it. With a smooth motion he looped it back into the air,

Al Was working with the skill and accuracy of a million catches. The heavy copper rod was floating anew. As the bar reached the apex of its journey, Al reached out and caught it again. Spinning on his left heel as he carried it horizontally around in a full circle, looped down, arched it up and let it go in the air again. As it floated, he moved with it but this time instead of grasping it in the jaws, he brought it to rest on the outside of the tongs holding it in a motionless balance. He turned his wrist slightly and slid it down to one end. With another quick wrist move he grasped the end of the bar and let it swing down like a door on a hinge. He tossed it again into the air but this time it floated toward Jason.

Jason reached out and caught the airborne rod. The heaviness surprised him. What a second ago had floated like a feather now had the weight of a sizeable hammer. Al handed the tongs back to Jason.

"When you're not chasing the crane, loading the furnace, or learning the size of the mills, practice with the tongs. We'll start you on the Hill next Monday."

Al turned and walked back towards the overhead office. "Another new one," he thought. How many had he seen come and go? Young bucks full of confidence and hate and anger and all the fine tools of youth. As he walked, he remembered back to the early days. Just the thought made him want to sit down with weariness. He glanced around to see if anyone was watching and slipped out through a side door.

It always amazed him. A step through the door and he was on the dock next to the deserted railroad spur. He sat down on an old reel and leaned against the red brick of the vibrating building. Here the morning sun warmed his old bones and the noise and hot haze of the factory stayed on the other side of the wall. He drifted back. Back to the first days of the rolling mill.

New mills, tall and shiny. His first sight of Big Bertha was thirty years ago. Bertha was the first mill the hot copper ingots saw as they rolled out of the gas-fired furnace. She was twenty-two feet tall. She held two horizontal stainless-steel rolls with vertical groves in them. It was much like the old wringer rolls in the original kick-start Maytag washing machine.

His mind raced back even further to a musty smelling basement in the farmhouse where he was born. He was only six years old and watching his mother kick down the washing machine foot pedal. The gasoline motor puttered to life and he immediately smelled the fumes from the leaky exhaust that fed out the cellar window.

His mother stood there in her plain cotton dress with fine tatting on the sleeves and around the neck. Her hair was held in place with a flowered scarf, tied like a bandana.

She brushed a wisp of hair away from her eyes, and poked the clothes down into the sudsy, slowly agitating water using a three-foot-long, broken broom stick. Light foamy suds clung to her slight forearms and a few bubbles floated above the washing machine and wringer.

She reached into the rinse basin and pulled out a pair of his father's bib overalls. A flip of the switch, and the wringer began to move. The two-white rubber-like horizontal rollers were utilized to squeeze the water from the saturated laundry. She fed the overalls into the wringer and the water squished and squeezed and formed rivulets as it found its way back to the rinse basin.

"Don't you ever get your fingers close to these rollers Allen! You get one finger caught in there and it'll gobble your whole body up. You'll come out the other side looking like those pancakes you ate this morning."

The vision of being pulled into those slowly moving rollers had given him nightmares. Then the first time he'd looked at the rolling mill, Big Bertha, he'd been in both awe and fear.

The mill whistle brought Al back to the present. A short and then a long slow blast, calling for the first bar of the new day. Al didn't need to watch to know exactly what would happen next. His old ears gave him the picture. He heard the exit door of the big furnace open. The hydraulic cylinders moved the pick-up arms into the fiery hell and drew back a cherry red ingot of copper.

As a hot bar was pulled out, a cold bar was pushed in and the entire circle of bars within the big doughnut furnace cycled like the carrousel on a slide projector.

From the furnace the bar went down a roller conveyor, into the giant set of horizontal rolls. Bertha groaned as she accepted the first bar. Lubricating water splattered and splashed as the bar was pulled though the first pass.

Jason was still wondering about the new tricks he'd just seen when in his mind he caught a glimpse of a young boy standing in front of an old washing machine. There was also a pretty young lady in a lace fringed cotton dress and the slight smell of exhaust fumes. He shook his head as if to clear it. "Get in the ball game," he thought. "Don't get caught in your daydreams back here or you'll end up fried or crushed."

Daydreaming had always been a problem for Jason, one he'd never been able to control. Visions would pop into his mind and off he'd go on an adventure. He was always getting himself into trouble. With teachers, with his parents, and now with his wife and bosses. No matter how he tried the visions would come from out of nowhere. By the time he returned there was often someone mad because he'd neglected the present. Many times, he felt he'd overcome the problem by learning to control it, but soon he'd slip away again.

Jake, the furnace man was moving the big steel cradle towards the stack of copper bars. He swung the one-man hoist into position and dropped the cradle onto the top layer of bars.

The cradle locked in with a metallic clang, separating six bars from the stack and lifted them towards the steel rollers that led to the furnace.

"This is all simple stuff," he was saying. "It's a continuous operation. The bars come into the bar dock by rail and are unloaded and stacked by a forklift. The overhead crane moves each stack to the furnace and the bars are loaded onto the furnace conveyor six at a time. The hydraulic rams push a cold one in and pulls a hot one out. I hold the furnace temperature at a constant twenty-one hundred degrees."

Jason thought back to his high school physics class about the melting point of solids. Copper, he remembered, melted at two thousand degrees. "Why doesn't it melt?" He asked.

"Ain't in there long enough. It's a continuous operation. If something goes wrong somewhere on the Hill, you need to crank the temperature back fast. Otherwise, you end up with a furnace full of liquid copper. Its only happened once. It took two weeks to jack hammer the big glob out of the bottom of the furnace. "

Jason looked down the long row of furnace controls. Each section had its own circular thermostat with a continuous red pen recording time and temperature. "Simple," he thought, "simple for Einstein maybe."

Jake continued, "The furnace indexes every eight seconds. The first set of rolls the hot bar sees is Big Bertha."

Jason watched the cherry red ingot rolling toward the giant mill and was awed by the force. The hot bar rammed into the entrance rolls and the big mill groaned under the strain. The pointed nose appeared out the back of the mill, smaller in the thickness but longer in length. The bar finished the first pass and came to rest on a long steel tilt table. The table tilted and the bar slid to a new conveyor with a dull clang. The powered conveyor carried the hot copper back toward the bar dock to the next set of mills. On it went, back and forth, each pass shaping the ductile bar into a longer and longer piece of copper rod. Each pass was automated. Each tilt table and each set of powered conveyor rolls was controlled by limit switches and stops.

Each automated station could be over-ridden by a manual control and the manual over-rides were all controlled by two pulpit operators; one next to the Super's office at the front of mill and the other adjacent to and above the rear part of the rolling mill…all automated, that is, except the Hill.

As the rod became smaller and smaller, it became much flimsier. The nose became black and distorted.

The elasticity, plasticity, ductility, and malleability of the copper varied with temperature and it could no longer be guided automatically into the smaller and smaller oval grooves in the stainless-steel rolls. The copper rod at the last four pairs of passes had to be handled manually with steel tongs. Jack swung the cradle back towards the stack of copper bars.

"Go up those steps over by number four mill and introduce yourself to the pulpit operator. Everyone calls him Sky. He'll explain how the mill runs. Oh yeah throw them tongs and that chunk of copper in the footlocker next to the control panel. You can practice later on catching that cold piece of rod. It's just like the hot stuff only a hundred feet shorter." Jason heard him laugh as he made his way towards the rear pulpit.

Catwalks everywhere. He walked above the tilt tables and conveyors, feeling the heat radiate through the soles of his shoes and the legs of his pants. The fiery metal fascinated him. Water, sizzling like spit in a hot skillet, danced over the hot ingots. The seemingly translucent bars passed beneath his worn greasy work boots. He stopped and watched as the bars rolled by. The noise and smell and new territory filled his senses. New experiences always fascinated him, and the new excitement of adventure keened his being to a sharp edge.

A shrill whistle blew just a few feet from his ear and both feet came off the catwalk at the same time. His grip on the handrail went to death squeeze and he felt his bowels almost let go. When his feet touched back to the catwalk he automatically went into a crouch and scanned for flying hot copper. He couldn't see anything unusual. "Son-of-a-bitch! Fuck!" He thought. He hated being scared more than anything. It was always the same, first fright and then anger. He looked back to the furnace, but Jake was still loading bars and hadn't seemed to notice. He scanned the area but couldn't see anyone. He continued shakily toward the stairs that led to the pulpit up above the rear of the giant mills. Jason opened the door and walked in. He stopped and allowed his eyes to adjust to the darkness.

Sky sat in a worn out, big armed, roller chair, all three hundred plus pounds of him. His jowls sagged on each side of his fat face and rolls of flab hung over his elbows. His thick fingers grasped a coffee cup which he carefully brought to his lips. A stream of black tobacco juice spewed into the brown stained cup from his big lips. Sky put the snus cup down and picked up a similar cup to take a long sip of coffee. He sat back and smiled a big brown toothy smile at Jason.

The control panel before Sky was seven feet long and two feet wide. It held a vast array of blinking red, green, and yellow lights. Three TV screens were mounted above the panel allowing him to see the blind side of every mill that wasn't visible from the three long tinted windows. Jason looked down to the catwalk where he had almost shit his pants then looked back to Sky. The fat man was barely able to contain his glee. Jason knew someone was fucking with him and he was fairly sure it was tubby here. He reminded himself of another of the factory rules: 'Train yourself to be dumb. You didn't hear anything. You didn't see anything. Therefore, nothing happened.' Sky looked away to the far end of the control panel to find a blinking red light. As he did Jason thought about switching his snus cup with his coffee cup.

"Don't do it," He cautioned himself. "Don't fuck with these guys. At least not yet."

Sky flipped a switch and the red light switched to green. The intercom light came on and the Super's voice shattered the silence.

"What the fuck was the whistle for?"

Sky pushed the talk button and spoke into the mike. "Nothing back here, Ken. Someone must have leaned on the whistle rope up there." There was a long pause and then a string of obscenities came over the speaker.

"Where's Patch?"

Sky pushed the button down again, "Who's patch?"

Jason opened his mouth to introduce himself but caught himself just in time. This was the tightest of all departments and he was sure they all knew he was coming on today. He kept his mouth shut.

The intercom came to life again. "Don't be fucking around fatso or I'll have your ass up to personnel quicker 'n you can spit." The line when dead.

"How's about giving me some more head, Morehead," He bellowed into the dead mike. Then he began to laugh. He laughed a belly laugh that was much too long-winded for the stupid pun. He laughed till the black tobacco spittle ran down his double chin. Sky was relieved at having something to laugh at. It was all he could do to keep a straight face when Jason had come through the door.

Seeing him jump and then duck on the catwalk had been the fucking funniest things he'd seen in weeks. He'd done it to about everyone new, but this one was special. A thoroughbred; quick as a cat and skittish as a new colt. It was a blessing to be able to laugh at something and not completely give away to his guilt. He spit into the proper cup and wiped his chin with his sleeve.

Not a new routine, Jason noticed. It was quiet again in the pulpit. Nothing but the rumble of giant steel mills which permeated his being. He watched with amazement at the matrix of events going on below him. As the copper left the last of the automated mills it headed in a straight line toward the last eight passes. Four pairs of small mills, not much taller than a man, stood on what the mill men called the Hill. The Hill wasn't actually a hill, in fact, it was only a little above the ground floor. What made it appear to be a hill was the fifteen-degree steel slides that sloped away from the mills and ran all the way to the basement.

In front of each pair of mills stood a man with a pair of tongs. With their backs to the Super's office and their asses almost touching the mills, they stood waiting for the hot snakes. Through the mill smoke Jason could see their forms. With their folded arms cradling a pair of tongs they waited. Through the haze he watched as the long cherry-red bars raced towards them.

Big John stood at the first station, with his arms folded. He waited. He was old for a sticker, as least fifty he reasoned. John wasn't sure. Numbers weren't exactly his cup of tea. He remembered prior to WWII.

The recruiter had asked him his age and birthdate. John had answered him with a question. "How old does I have to be to join up?" The recruiter told him, and John asked, "If I is that old, then when was I borned?" With the next answer his age was set. It had gone pretty much the same way at the personal office in the copper factory.

"How old are you?" The personnel manager Bill Confer asked.

"Well," he answered, "I was sixteen when the Big War started. That makes me as old as I am now." The manager wrote a few words and asked him to sign.

Signing was always a bit embarrassing for John. He signed JC in a half print half scroll his mother had shown him. She'd opened the Bible and pointed out that the first letters in his first and last name were the same as Jesus's name. All he had to do was copy from the Bible. That had been his first and last lesson in reading and writing.

"Where is you now, John Curry?" he asked himself as he watched the long red snake racing toward him. He'd been asking himself the same question all his life. First in the fields, then at Normandy and now, again in the copper mill. "Why's you here?" He would ask himself, but he never came up with an answer. He'd go ahead and do whatever he had to do. As long as the sun came up in the morning, he'd do it all again.

The company had started John in the wire mill. The wire was too small for his big fingers. No matter how hard he tried he could never get the wire into that little bitty hole in the die.

He remembered the day he'd become so frustrated with the equipment that he'd gone into a rage. He'd pounded on the steel cover of the machine so hard the metal had split, and the rivets had popped. They might have fired him over that incident, but the company couldn't find a foreman with enough courage to give him the bad news. They'd decided to move him to the rolling mill, and he'd been there ever since.

He'd fallen in love with the place. As a crane chaser he learned the proper way to pick and sit a load. From the crane hung heavy steel chains that a man could get a hold of.

The mills were held together with two-inch-wide nuts and bolts. The open-end wrenches weighed as much as sledgehammers and the mill caps weighed two hundred pounds.

Some of the old timers claimed that when John was young, he could lift the caps off without calling for the overhead crane. The wire back here wasn't wire at all, but copper rod and the machines wouldn't eat your fingers. He shuddered at the thought. Sure, he wore the burn scars from being kissed too many times by the hot rod, but at least he had all his fingers. The scars he wore with honor and back here, he got respect.

The rushing nose of the hot rod hit the first manual mill with a dull thud. John counted two seconds and reached down by his left ankle with his tongs and grabbed at the nothingness. It always amazed him. He reached with open-jawed tongs into the stream of cooling water flowing from the mill guide and after he closed them, he always came up with a hot black nose of copper rod. That had been the hardest for him in the learning. "How's you know when to grabs it?" he had asked.

Old Al had been patient with him. "Just count to two after you hear the thud. Reach down and grab and you'll come up with the bar in your tongs."

John hadn't believed Al. The first bar he tried to catch he'd waited until he saw the nose come out of the mill. By the time he grabbed it the nose was ten feet away and headed for the basement.

"That's called catching it long. You can't stick it in the mill unless you catch it close to the head."

Whenever a new man caught one long, or missed it completely, the excitement would begin. Most men never forgot their first bar and John was no exception.

There he stood, his tongs gripping a nine-sixteenth oval snake of cherry-red copper ten feet from the nose. The nose drooped down on one side of his tongs and the remainder of the copper rod was spewing out of the mill guide, running down the steel side toward the basement.

"Drop it!" Al yelled over the noise of the rumbling mills. Instead of opening his tongs and letting go of the rod John had let go of everything. His tongs clattered and clanged their way down the greasy steel slope. "Get behind me!"

John moved behind Al, which was not easy. There was little room for one man at the first station and two men together was exceptionally tight. Two was almost impossible with the size of this African.

Two seconds spent catching it long, another two seconds getting him to let go and three seconds getting into position to cut his way out of trouble. That left nine seconds to get back in the game. Al reached with his tongs and grabbed the rod.

The weight tested his strength and pulled at his joints. He hated the first station. Heavy and hot and no matter how much skill you had you couldn't get away from the weight.

Catching the bar in the middle overloaded his light frame and took all his strength to drag the load back. He dragged the bar back and over the two-foot-high steel pedestal called the 'dead man'. His hand axe lay on the pedestal. Holding the tongs in his left hand, he picked up the axe with his right and tapped the hot copper flat to the malleable steel. He raised the axe high and with a strong swing he chopped a new nose on the copper rod.

He released the axe. With both hands back on the tongs he released pressure on the bar and slid the jaws of the tongs up the rod to within ten inches of the newly cut nose. As he spun towards the receiving mill, he noticed the next bar racing towards him. He jammed the cut end into the waiting rolls and heard the thud of the next bar hit.

He spun, reached, grabbed and came up with the next bar. The point at which the bar was grabbed always had a slight bend in it and as Al came up with it, he released and slid his tongs up closer to the head, thereby demonstrating the art of the double catch. He stuck the bar into the waiting rolls and relaxed with the luxury of having a full fourteen seconds before the next bar arrived. The hot copper spewed out the left side and fed into the rolls on the right side. The loop between the two mills grew away from the stickers and towards the basement. A long straight piece of scrap, from the bar Al had just cut, lay on the sloping hill. He reached for the cooling piece of slowly blackening scrap with his tongs.

Grasping the near end, he raised it high above his head and cracked it like he was cracking a whip. The rod raised up, folded over and with a flick of his wrist, the entire piece of scrap floated to the aisle between the inclined steel slopes. It hit and lay sizzling in the dirt and grease.

One of the foremen walked over and began rolling the piece of scrap into a tight tangled knot. His tallow-laden gloves smoked and caught fire as he worked. He slapped the gloves together knocking out the fire and continued working.

Al caught the next bar, stuck it and reached for a spare set of tongs he kept on the top of mill.

"Next time drop the copper but hang onto the tongs," He said.

Big John wasn't awed by the display of skill, he'd seen stranger things in the war. He wasn't scared by the heat and fire and red metal. He'd dug sweet potatoes all day in the hot sun and this wasn't much hotter than that. He was mad. He was mad and embarrassed because he hadn't believed Al when he said he couldn't wait to see it before he caught it. He was mad because he'd dropped his new tongs and they were laying on the bottom of the long sloping steel pit.

"Get outs my way," He said. Al had never left anyone alone after one bar and he didn't want to step aside now. He also didn't want to stay next to anyone showing this much fury. Al stepped aside.

John gripped the cold steel handles of the tongs and watched the next bar hit. "Onnne, twooo," he counted out loud. He reached down into the cutting water and came up with his tongs full of hot copper. Just like catching a big black snake in the swamp he thought. The force at which he'd grabbed it bent the nose of the bar into a U shape.

Not knowing how to perform a double catch he stopped at the dead man and pushed the nose down onto the soft steel to straighten it. He picked it up, continued his turn and slammed it into one of the empty four holes at the entrance of the waiting mills.

The rolls caught the bar before he could turn it loose. The quick jerk of the bar slammed his fingers into the mill guard. He released pressure on the tongs with a grimace and spun to catch the next bar. It wasn't even in sight yet.

Al was amazed. "You got to release it as you stick it," Al hollered.

"No shit!" John thought, as he felt the pain course through his hands. He caught sixteen in a row, straightening each one on the dead-man and slamming it into the rolls. The first one he missed created a shutdown because Al couldn't cut and re-stick the bar for him. Each shutdown cost the company five bars of scrap, plus whatever extra work they all had to do to get going again. No one complained, at least not to john. It took Al a week to teach him some of the fine points but from then on, the first station belonged to him. John loved the first station, and it was the only one he'd stick at. The other stations were faster and flimsier, and he couldn't seem to master the finer points of the double catch. John knew they were getting a new man today. He also knew that next week, he'd be the one to break him in. "A lot of extra work," he thought, but it went with the job.

Chapter 3

From the rear pulpit area Jason had watched the mill men work. Sky sent him back to Jake and then to the basement to pull scrap. With each job came new experiences. The week raced by and the weekend was upon him before he knew it.

Jason picked up his paycheck from the Superintendent and headed for the time clock. He went home for a quick sandwich and then out for his Friday evening at the bar.

He pulled his car onto the cracked sloping blacktop next to the Old Mill Tavern. Jason listened to the river on its never-ending way to somewhere south and suddenly thought of LeRoi. Would he be there? Did he really exist? He felt his front pocket assuring himself that his paycheck was secure and walked through the back door past the foul-smelling restroom into the sounds of the old bar.

"Evening, Patch," the bartender said, as he popped the top of Jason's favorite beer and reached for his paycheck. Jason signed it and slid it across the bar, Frank cashed it and handed back a pile of money, less the beer. Jason picked up the beer and walked around the partition that partially separated the large room. Tables and bar stools on one side, pool table, shuffleboard and pinball on the other. Near the back of the room against the wall sat the dark-haired man sipping from a shot glass.

LeRoi looked up and made a slight toast of greeting. Jason walked over and sat down. "Good to see you, Jason. I didn't know if you'd show or not. Are you ready for another lesson in time travel?"

Jason was sober this time and he wanted to be sure about what was happening. He was afraid someone was playing a practical joke on him. "Last week you said you could teach me to travel in time. You were joking, right?"

"No, I wasn't joking. Don't you remember the old playground? The way it felt, the way it tasted, the way it was. Those things didn't come from my mind, they came from yours. Can you forsake truth?"

"That wasn't truth, that was just a daydream and those always get me in trouble." Jason wanted to believe, but everything he'd ever learned denied the possibility. Still, he needed desperately to believe in something.

His life and relationships were typical of others of his generation. Externally he appeared to be living the American dream but internally his life was a mess. Jason often felt as if he were in a vast ocean, drowning in the turmoil. He wanted desperately to reach out and grab a branch, or a life preserver of some sort.

LeRoi continued as if Jason's reservations and objections were dust to be blown away with a slight breath, "Do you remember last week's lesson?"

"Sure, the earth was born and has a birthday every year, just as I have a birthday every year. Real Time is marked by those birthdays and is recorded by clocks and calendars. Each minute is a minute unto itself and cannot happen again."

"That's great. You get an "A" for your first test on time travel. That's all there is to that lesson."

Jason liked this. He could count the A's he'd received in school on one hand. They hadn't come nearly this easy.

People were steadily filing into the bar. Music was beginning on the jukebox and the first pool game of the night was starting. Quarters were already lined up four deep on the edge of the pool table.

Jason excused himself and walked to the table to put up his quarter.

LeRoi waited for him to return. This was going to be difficult, he thought. Jason had obviously never learned that the most valuable gift you can give a teacher is your attention.

He also was not taking this too seriously. "Not to worry," he thought. If I'm a good teacher, I can overcome petty distractions. With the proper routine I can capture his mind so totally that the quarter will never be seen again. LeRoi watched as Jason walked back to the booth.

"Okay, I'm through the first lesson. How many lessons are there?"

"Four."

"Are they all this easy, and this short?"

"Yes", LeRoi replied.

"And if I get through all the lessons, I'll be able to travel through time?" asked Jason.

"Yes." LeRoi replied.

Jason was interested but his mind would not allow him to accept the possibility. The last time they talked had been like a journey. Jason thought the beer may have been the biggest factor in their traveling, but now he felt himself slipping under a spell. This new feeling couldn't be explained away by alcohol. It was like something he hadn't felt in a long time. It was like the way he remembered feeling about his first-grade teacher. Was it respect or friendship or intense interest? He couldn't quite put his finger on it.

What the hell! This LeRoi guy acts like time travel is possible. Jason didn't have anything planned except drinking himself into oblivion anyway.

"Ok, I'm ready for lesson number two," he said, as he slid into the booth.

"Have you ever played baseball?"

"Have I ever played baseball? Does a bear shit in the woods? Is the sky blue?"

Jason knew everything there was to know about baseball. At least he thought he knew.

"I started playing when I was about five, playing catch with my Dad. I played park league ball, little league, pony league and high school. I still play softball. I can't think of anything I know better."

LeRoi saw the sparkle in his eyes and listened to the statement. He knew he was beginning to capture him. "That's great Jason, because today's lesson is going to be about baseball."

"Piece of cake," Jason thought, "Here comes 'A' number fucking two. "

"There are only two types of time. Real Time which you already know about, and Mind Time. Real Time is a constant and Mind Time is a constant most of the time.

Jason remembered about constants and variables from junior high school algebra. The rotation of the earth didn't change and so was a constant, but something which changed,

even occasionally, was a variable.

"You're saying time consists of two parts, one constant and one variable?"

LeRoi smiled inwardly. He had him now. The trap was closing, and the external influences were disappearing.

"Have you ever known anyone who could tell what time it was without a clock?"

"I can do that, except when I drink too much," He thought back to all the times he had used the gift to amaze his friends. He didn't know how it worked or understand how he could wake up exactly on the minute to go fishing. But when it was time to go to work, he needed two alarms and a nagging wife to get him out the door.

"How do you do it, Jason? If it happens, there must be a reason."

Jason didn't know the answer and LeRoi continued.

"Everyone has an internal clock in their mind. Everyone knows what time it is; some just don't have access to their Real Time file. It's there, they just have not learned to access it. Most of the Mind Time clock is synchronized with Real Time. Real Time is going along one minute at a time. One minute of Real Time is synchronized with each minute of Mind Time. If anyone asks you what time it is, you can tell them without looking at a clock. Most of the time you can make a close guess, and sometimes you're closer than others."

Jason was becoming bored. "What the fuck does this have to do with baseball?" he thought.

LeRoi was watching him closely. Any sign of his student slipping away clued him to ease up on the specifics and return to the lighter areas. It was like playing a fish. Just the right amount of tension on the line. Go too fast and you break the line. Too slow and the fish shakes the hook and swims off to other topics. There was no bond yet; no mutual respect; no trust. It was just a teacher and a student.

"Have you ever wondered how a batter hits a fast ball?" LeRoi continued without waiting for an answer. "Let's pretend you're in the batter box at Wrigley Field."

Jason's eyes brightened a little as he slipped into the fantasy. "The sun is high in the sky and you grip the bat tightly. You can feel a trickle of sweat running down your forehead. You step out of the box and wipe your brow with your sleeve. You step back to the plate and dig in with your back foot. You look to the pitcher sixty-one feet away. He grips the ball and looks you in the eye."

Jason felt his shoulders tighten. He could feel the heat and taste the sweat.

"The pitcher goes into his wind up and as his arm rotates forward you see the small sphere being released. This guy throws a good fast ball, and it blows by you at ninety-three miles an hour. 'Strike one.' The umpire says."

"No fucking shit," Jason thought. You'd have to have brass balls just to stand there, let alone hit the som-bitch,

"So why didn't you hit the ball?"

"It was going too fast."

"The good hitters hit it. They hit it clear out of the park. Have you ever heard their explanation of how they do it?"

Jason had to think about it for a while. What did they say? It was always the same.

"They say when they're hitting well, they can see the rotation of the ball. They claim they can see the stitching on the ball as it hits the bat. They say it's as big as a grapefruit."

"Come on Jason, do you believe that? Have you ever been able to see a fast-ball like that?"

"If I could see the fucking ball like that, I'd be out there playing ball instead of working at that factory. No, I've never seen the ball like that."

"Do you think they're all lying? Do you think they made up the same story?"

"No."

"Good, neither do I. Now let's try to figure out this riddle." LeRoi pulled a calculator out from his shirt pocket and began punching in numbers. He mumbled as his fingers played the keyboard.

"Ninety-three, times fifty-two eighty over sixty, over sixty-one and invert… equals a bit less than half a second.

"So, from the time the pitcher lets it go until the time the catcher catches it, you have a little less than half a second to react and hit the ball. That means that the answer to why you can't hit it, isn't that it's coming too fast, it's that you're reacting too slow. After all, the good hitters can hit it, so it is possible!"

Jason thought for a while. Not only could he not see the stitching or the rotation of a good fast ball, but he also always felt lucky to get out of the way of an inside pitch.

LeRoi continued, "Now, you're still at the plate and the count is 0-one. You need to use your imagination a bit but bear with me.

"I jump into my super-charged model-A street rod. I stop at home plate to pick up your mind and your eyes. We'll leave your body standing at home plate."

Jason smiled inwardly. Part of this lesson involved more daydreaming. He imagined his brain with two eyeballs somehow levitating above the passenger seat of an old street rod. He imagined at his command, his brain and eyes rotating three hundred sixty degree to see a panoramic view of the baseball stadium.

LeRoi popped the clutch and they sped toward the third baseman, heading for the outfield. They passed the left fielder and turned right at the warning track. The ivy-covered walls of the old stadium stood in sharp contrast to the street rod driving around in the somehow out of place, disjointed experience of the ballpark.

They turned right again at the center fielder, gunned it going toward the second baseman and hit fourth gear.

Jason rotated his eyes towards the dashboard and glanced at the speedometer. He noticed they were steady at ninety-three miles-per-hour.

They blew by the pitcher just as he released the ball. The ball and car were traveling at exactly the same speed, both flying toward the motionless body at home plate. The body standing at home plate belonged to him, but the eyes were missing.

"Now look at the ball!" LeRoi shouted.

Jason looked at the ball as it floated outside the window. With the ball and the car both going the same speed, it was as if it were suspended with no forward motion. It was rotating from the spin the pitcher had put on it.

"I can see, the rotation. I can see the stitching on the ball." The car and the ball blew by the batter at the same speed. The ball banging into the catcher's glove as the street rod slid sideways and came to rest at the back stop.

Jason heard the umpire call "Strike two!"

What a student, thought LeRoi. Just get him going on the correct path and he will explain it to himself. "You could see the ball just like the good hitters can?"

"Sure, but we were traveling at ninety-three fucking miles an hour."

"But if you were traveling that fast, you could hit the ball."

"Sure, I could hit it if I could see it like that," he answered.

"That's the way the good hitters hit it. They don't do it with a car, they speed up their mind and Real Time appears to slow down, thereby allowing the reflexes within their body to react and hit the ball."

"I don't understand."

"It's simple. The hitter's mind kicks into high gear and Real Time appears to slow down. The ball appears to be coming much slower than it really is, allowing them to react and hit it."

Jason's mind was working now. He wanted to go out and try it. He slipped back into his imagination and felt himself back on the ball field. The count was 0-two. LeRoi saw Jason's eyes partially glaze and knew he was back in his imagination. He'd lost Jason's train of thought.

Jason stepped back out of the batter's box and knocked the dirt off his cleats. The crowd was anxious and noisy.

Jason wiped his brow again and stepped back in. He looked to the pitcher and they locked eyes, both concentrating on the other.

Now the pitcher holds the ball, and now he lets it go.

Jason saw the ball leave the pitcher's hand, saw it spin ever closer, he could see the stitches in the seams.

The power came from Jason's back foot. It traveled up his legs, through his back and shoulders and out his arms, wrists and hands to the bat. With a loud shout and a mighty swing, he pounded it out of the park.

Jason's sudden yell pierced through the random noise of the bar and caused both of LeRoi's knees to slam into the underside of the table. The bar became quiet and everyone turned their way.

Jason looked around and felt that old embarrassment that he had so often felt before. Caught in a daydream, not distinguishing reality from fantasy, he felt his forehead. This time he wiped away real sweat.

"What the hell was that for?" LeRoi asked.

"I just hit the fucking ball out of the park," he whispered.

The bar returned to its previous level of noise and confusion. Jason felt tired and was still embarrassed at his lack of control.

"Do you think I could really slow time down in a real-life situation?"

"You're not actually slowing down time. You're speeding up sensory input and Real Time appears to slow down. Some people appear to control it better than others. Remember Real Time is a constant. When it's past it cannot happen again. You can revisit it, but you can't re-do it.

"Yes, but the ball field I just visited wasn't in my past. How can I go somewhere I've never even been before?"

"There appears to be three regions that can be visited, the past, the, future, and the imagination. For some reason there seems to be space allotted for imagination. You can create an imaginary adventure just for the fun of it, for experimental testing of ideas, or for any number of other reasons."

Jason started to get that confused look again. He felt he was just on the edge of understanding.

"Have you ever heard of the Canadian poet, Gordon Lightfoot?"

"You mean the singer, the one that sang about the Edmond Fitzgerald?"

"That's right. In that song he has an excellent example of what we're talking about."

Jason's mind backtracked through the song, but he didn't find anything on baseball or time travel. "How's that?", he asked.

"Do you remember the verse, 'Does anyone know where the love of God goes, when the waves turn the minutes to hours?'"

"Yes."

"He is explaining the same phenomenon. Real Time is going along, and one-minute Real Time equals one-minute Mind Time, and then all of sudden he's saying the waves have made one minute equal one hour.

"Have you ever fallen into extremely cold water?"

"Once a couple of years ago I was canoeing the Flat Rock River east of Indianapolis. There was ice along the banks. The water was fast and rolling. A submerged log, a missed stroke and a bad weight shift put my partner and me in the frigid water. The icy water wasn't too bad on my arms and legs but when it hit my nuts, I almost passed out. I remember wishing I could levitate myself out of the cold fluid and up into the warm sunshine. The shore was only twenty feet away, but it seemed like hours getting there."

As he thought back Jason realized he could swim twenty feet in less than a minute. At the time it had seemed to take him forever.

"Yeah," He said, "I've been in really cold water before."

LeRoi had been sitting there watching him. He enjoyed getting all the details of Jason's' experiences.

A pretty woman walked by the table almost brushing Jason's arm. Much too close. Jason looked up and followed her movement across the room with his eyes. He looked to the pool table and realized it was past his turn to play. They were finishing a game.

He picked up his glass and realized he was only half-way through his first beer, and it was warm. He glanced at the clock and knew that the night was passing him by.

He got up and walked to the pool table, put his quarter in the slot and listened as the fifteen balls rolled to the waiting collection area at the end of the table, as he knelt and reached for the balls he glanced at the booth he'd just left. It was empty.

LeRoi was already slipping down the bank towards the river. "A good night," he thought. Good to be out of the smoke and noise, and stupidity. Good to be welcomed by the vastness of the night. He drank in the stars and openness as he and the canoe slid into the rolling current.

Back at the bar Jason's arm went back and came forward, pounding the cue stick into the solid white ball. The object balls exploded and scattered to random locations across the table.

Chapter 4

The pool game hadn't gone well, Jason's mind kept wandering and he'd lost without ever being in the game. Rude of LeRoi to just run off without saying good-bye. He'd tried to catch a buzz, but the beer wouldn't take hold and he slipped into another daydream.

It was three AM and Frank was calling for ashtrays.

Jason's Friday was finished. He'd driven home and climbed into bed as always, but his dreams had beaten up on him.

It was now five AM Saturday morning. He sat in his favorite chair in the living room trying to analyze what had happened in his dreams.

He had dreams, just like everyone had dreams and after he woke up, he tried to figure them out. Usually, he traced them to something he'd seen the previous day or to something someone had said. Occasionally he'd have a dandy, one he couldn't figure out. This one had been more than a dandy.

In his dream he had been sitting on a bluff overlooking a river. All his brothers and sisters were sitting around him. His father was sitting across from him sipping a cup of coffee.

"Thanks for the coffee son," his father said.

A warmth flooded over Jason. He was so happy to see his father. It had been so long. Why had it been so long? Then his memory slapped him hard.

"It's been so long because your dad died eight years ago," his rational mind told him.

As soon as the thought appeared, the scene started to fade.

"Don't go," He thought, "I want to ask you so much! No, I won't ask you anything; I just want to be near you for another second."

But it was too late.

It was always the same. Some scenes that didn't make sense. A two- or three-word message that didn't mean anything and then his real-world mind would destroy it like an idiot with a ball bat pounding on a glass Madonna.

So, there he sat, with loneliness crushing in on him. He sat in the darkness, tears wet in his eyes and alone. What did the coffee have to do with anything?

When he was young, his father was on third shift; eleven at night till seven in the morning. Jason's only job had been to make coffee for his father and fill his thermos. Was that what the visitation was for? Just to say thanks for the coffee. How could such a little thing be so important? Why couldn't he stick around for a chat?

"Next time," Jason thought to himself, "I'm going to reach out and grab him. I'm going to grab both his upper arms in my hands and hold him and I'm not going to let my memory fuck it up."

Jason felt the warm sun on his face and opened his eyes. There stood his wife, arms folded, giving him that 'you're crazy' look. Jason sat, naked, sleeping in a sitting position in his favorite chair. He got up as calmly as possible, like this was the most ordinary thing in the world and walked quickly to the bathroom.

The weekend seemed muddled, like nothing was fitting into place. He read in the paper about a man wo had been attacked by a mugger in the park.

The paper quoted the man saying, "I saw him coming out the corner of my eye, and all of a sudden it was as if he was moving in slow motion." The guy had beaten the mugger half to death.

Then there had been the parade on TV. A team of horses had bolted into a crowd and dragged a female pedestrian half a block. It had only lasted an instant but while she was on the stretcher being loaded into the ambulance, she had claimed the dragging had lasted forever.

It all made sense to Jason now. It was as if a light had been turned on in his head. In times of severe stress or in a life-threatening situation, the mind seemed to be kicking into high gear, giving the body more time to survive.

It seemed so simple and obvious that he wondered why he hadn't thought of it before. Now everywhere he looked it was happening. Mind Time would speed up and real time would seem to slow down. It was like learning a new word and recognizing it every time he heard it again.

He tried to explain it to one of his friends, but he had come up with so many wild ideas in the past that his friends humored him and acted like they understood. He knew they didn't. He lay in the grass in the sunshine and told his son the whole story. His son listened and heard it all. His son understood as well as his friends had, but it was much more fun with Brian.

"The mind speeds up and Real Time appears to slow down."

"Yes, Dad. Look at this grasshopper. How does he jump so high? Why does he spit brown juice when I catch him? Why do bees sting?"

At least with his son he didn't feel all the frustration. He could lay back and say anything he wanted, and the day would fly by. He avoided thinking about the factory. He pushed the rolling mill out of his mind and tried desperately to enjoy this moment with his son. Somewhere in the back of his mind a voice was telling him it wouldn't last forever.

The thought filled him with so much dread that Jason forced it quickly from his mind. He had been getting better at ignoring reality. "Just live for the moment," he thought. "Don't let the real world fuck up this day."

Chapter 5

Monday Jason was early for work, something that was a rarity. Everything felt cold to Jason this morning. His work boots were stiff and cool. His blacksmith tongs felt heavy and his clothes were stiff and scratchy.

Today he would attempt to catch his first bar of hot copper. He had heard the stories about Big John, how he' d learned the job after only missing one bar. Others had taken weeks to learn and some never had.

He had practiced with the short piece of copper, each time imagining himself catching the hot bar 'from the exit guide at his left ankle and sticking it in the entrance guide to his right. He had tried some of the tricks Al had demonstrated but was seldom able to duplicate the most rudimentary moves. He desperately wanted to slow down the clock.

The more he wanted to avoid that first bar the faster the second hand seemed to go. He imagined himself catching the first bar, sticking it, and never missing any. He imagined the rest of the mill hands talking about him as the only one to ever learn on the first bar. His imagination allowed him to be the best from the first catch.

His reality mind was screaming for him to run. He walked back to the furnace area and watched Jake loading bars into the fiery hell.

"Morning, Jake."

"Hey Patch, how are you doing this morning?"

"OK. I'm going to the Hill today," he said, like it was something everyone didn't already know. "How about explaining the whistle to me?"

"Not much to it. There's a long rope that runs above the mills within easy reach of the stickers. If there is a problem, the mill men swing their tongs back and hit the rope. One short blast means emergency stop. Two short blasts mean bring 'em on full speed. One short followed by a long slow whistle means bring 'em on half speed."

Jason thought back to the first day he'd walked across the catwalk. "When I was on the catwalk the other day, the whistle blew, and the mill didn't shut down. Why not?"

Jake smiled, "That's called an inadvertent whistle, a mistake."

"How do you know a mistake whistle from the real thing?"

Jake hesitated as if searching for the right words, "You should be talking to Big John. He'll be get's up there 'bout now."

Jason headed for the Hill. He knew the answer to his question. Fatso Sky had blown the whistle as he walked across the catwalk. What bothered him was that the whole crew must have been in on it, or the mill would have been shut down. Back here the crew worked together. It took everyone working together to get the copper running and it also took everyone working in concert to shut down properly.

If everything weren't done in the proper manner scrap increased and old Super Ken Morehead would have their asses. Jason already suspected a conspiracy.

"Fuck em. They're all a bunch of stupid fucks anyway," he thought, but deep down he wanted to be accepted. He wanted to be a part of this group. He wanted desperately to be a part of something. "Don't let your paranoia fuck this job up," he thought, "You're just anxious about your first day on the Hill. Others have done it, and you can do it." No matter how he tried to talk himself up, the truth was, he was scared.

He feared being burned, scared of embarrassing himself, and scared to death of reality.

John was standing at his station. His arms were folded cradling his tongs and waiting. Jason walked down the aisle and turned at the catwalk that traversed the steel slides that carried the copper. He turned at the aisle way that led to the sticker's station and walked towards John

"You be Patch I reckon," John said, holding out his hand.

Jason took his hand, "Jason Strider's the name."

John shook Jason's hand and felt the strength. He'll need it all he thought, that and more.

"Yous just stand out there between the stations and watch what I do. After a while, I'll let you get in here and you can catch one."

Jason was taking it all in. John looked to the forward pulpit area and nodded. Joe, the forward pulpit operator, buzzed back to Sky and asked for an all clear.

Sky looked to Jake and hit the furnace buzzer. Jake looked to the darkened rear pulpit and, although he could see nothing but dark glass, he gave the thumbs up. With a slow rumble the giant mills began to live. Cooling water came splashing out of the mill guides. Huge stainless-steel grooved rollers powered by electric motors and coupled with brass couplers, vibrated up to speed. The entire building took on the hum and vibration of a living entity.

John waited and listened. When all the sounds became familiar, he reached back for the whistle rope with his tongs. A short blast followed by a long steady whine, and Jason heard the door of the furnace open.

In his mind he saw the hydraulic ram reach into the fire and lift the first bar of the day to the waiting conveyer. He heard Big Bertha groan as the hot ingot met the first rolls and saw the cloud of water vapor rise into the high ceiling haze.

"This is easy stuff," Big John was saying. "You waits till you see the nose hit the intake roll. You counts two seconds and then yous reaches down and catches it." John no longer needed to count. He'd caught so many bars, his timing, so automatic, he couldn't miss if he tried. He could turn his back on the bar, listen for the thump and still turn and catch it.

Jason saw the bar racing toward them. It hit the rolls with a dull thud.

"Onnne, twooo." John reached into the cutting water streaming out of the bar guides and came up with the nose of a long hot copper bar trapped in the jaws of his tongs. As he turned, he stopped at the knee-high pedestal in front of him and straightened the bar. The remainder of the long piece of copper rod was spewing out of the exit hole next to his left ankle and the middle loop was racing for the basement.

John held the blackening end of the next bar close to Jason. "That's what she looks like up close. Yous just grabs it and Stick it"" He said as he stuck the nose in the waiting hole next to his right ankle.

The bar jerked as the waiting rolls caught hold. The loop that had been racing for the cellar slowed its downward speed as it spewed out of one hole and was gobbled up by the other. The sticker at the next station reached for the bar, caught it, double-caught it and passed it to the

next sticker. The finish sticker in turn passed it into the finish rolls just as the tail was exiting John's station. The bar was like a skier going through a slalom course, the men being like the pylons. It wove in giant red S's as the men caught and guided it on its way.

The first bar, after being coiled, dropped into the water with a sizzling hiss just as John was reaching for the third bar of the day. He reached, caught, straightened it on the deadman and stuck the third bar.

"Get in here and catch the next one."

Jason turned red, "But… but"

"Nows as good a time as any. Get on in here!"

Jason stepped over the hot rolling copper. It was close quarters. He could feel the presence of the big man behind him. He quickly glanced around but no one else seemed to be watching.

In fact, the entire crew was watching. Both pulpit operators were on their feet with their hands on the controls. Ken Morehead was on his feet looking out the darkened glass of his office. Jake, the furnace operator had climbed up on the stack of cold bars. Old Al was standing on the catwalk behind Big Bertha, squinting through the blue smoke of the giant mills. Each wanted to see the young blond catch his first bar.

Jason watched it come. He hadn't expected it to happen so fast. He thought he could watch thirty or forty bars and then maybe step in and try one and then step out. No time to worry now he thought to himself, just catch the fucking bar.

It hit with a thud and he counted. One, two reach and grab.

He reached into the water but felt nothing. He hesitated, not knowing what to do and then felt the full force of the bar hit the hinge of his tongs. The bar pounded the tongs out of his hands and sent them rattling to the cellar. His first response was to go after them. He made a step towards the greasy steel downslope.

He felt a big hand grasp the back of his belt and heard a gentle voice say, "Step behind me."

He quickly stepped back, not that he had much choice. The strength that pulled him back couldn't be denied.

John stepped forward and reached for the angry hot metal. The nose was already twenty feet away and the longer it ran the tougher it would be to pull it back.

He closed his tongs, and the old worn jaws held the fast-running bar. He pulled the hot copper back to and across the deadman.

Holding the bar with his tongs in his left hand, he grabbed his hand axe and tapped the copper tight to the deadman. He made a quick, clean cut. He picked up the newly cut nose and stuck it to the waiting rolls. He turned with the skill and grace of a dancer and made the next catch. "Step out and roll up that scrap. There are gloves on top the mill and another pair of tongs."

Jason stepped out and reached for the gloves. John reached over and grabbed the twenty-foot-long piece of scrap with his tongs. He gave it a flip. It rose up out of the steel slide and came to rest smoldering in the aisle.

"We role our own scrap. If you let it get cold, it gets stiff and it's harder to roll." John reached and caught the next bar and Jason rolled up the hot scrap.

The rest of the crew had gone about their work. Not too impressive. Typical rookie mistake. Moved too fast and lost his tongs. They'd seen it all before.

The next bar Jason waited more than two seconds and caught it long. The one after he barely caught the nose.

Each mistake made the big black man sweat a little more and each miss made the scrap ball grow larger. It was a long day for everyone.

The week dragged on but by Friday Jason was catching most of the bars at full speed.

Chapter 6

Jason was early to the bar room Friday. He wanted to tell LeRoi about his new job and he had a lot of questions to ask. The parking lot was empty but for the old Mercedes belonging to Frank the bartender. Jason figured he'd just have to wait for LeRoi to get there.

"Evening, Frank," Jason said as he slid his paycheck across the bar.

"Evening Patch, Frank replied as he returned the pile of money and a beer. Your buddy's been here for about half an hour."

"You mean he's here? Wonder where he's parked his car? Do you know anything about him?"

"No, thought he was a friend of yours. First time I remember seeing him was with you. Prefers pissing in the bushes to restrooms is all I know about him."

Frank also noticed he didn't drink much and when the two were together neither one drank much. Frank didn't care for that. He also wished since they were the only two in the bar they would sit where he could see them.

Jason walked back to his learning booth, sat down and rushed into his dialog. "You ought to see what I'm doing at work. These long hot copper bars come out of a hole and I catch it with a pair of tongs. I saw in the paper where a guy was about to get mugged and Real Time slowed down. There was this lady that was run over by some horses and it only took a few seconds. She thought it took forever. If you speed up the mind and Real Time appears to slow down- if you slow down the mind wouldn't Real Time appear to speed up? If you can make Real Time slow down, couldn't you live a lot longer and…"

LeRoi cut in. "Hold on Jason. Slow down." This was almost as bas bad as him not paying attention. Too many subjects taken too fast would leave him just as empty as before, with the same chance for mistakes and misunderstandings. "Let's go over what we covered last week."

Jason had to think. Had it only been a week? Last Friday seemed a long time ago. Prior to meeting LeRoi his life had seemed simple. Now, the complexities were beginning to confuse him. He forced himself to focus on the last lesson. Baseball. A good hitter has the ability to slow down a baseball or at least make it appear to slow down, so he has a better chance to pound it out of the park. Some people appear able to speed up their mind and Real Time appears to slow down.

"It also appears to me that it may be a survival mechanism which kicks in automatically when you're in a life-threatening situation."

LeRoi liked this. His student was putting two and two together and coming up with four. He was starting with a basic truth and expanding it to other applications. He needed a lot of guidance but at least he had a chance to teach him the basics. "That's correct, Jason. That's 'A' number two."

Somehow the letter grade didn't seem to mean much this time. Somehow the excitement of learning seemed to overshadow the imaginary report card. "I've heard that just before a person dies, his whole life flashes before his eyes. Is that the same mechanism we've been talking about?"

LeRoi had to be careful here. He wanted to give definitive answers, as if he knew exactly what each of these phenomena were. But some of the questions could only be answered with theory. If later the theory proved wrong, he'd lose creditability and this student might tend to discount the more important parts. He pressed forward with the truth.

"Actually, no one knows exactly what happens before one dies but it's one explanation of that particular phenomenon. The mind does a desperate scan of all your life's experiences searching for an escape from the desperate situation you're in. So, in effect you could re-live all those experiences in a flash. Your mind would be processing data from all its data files at an extremely high rate of speed. In that instance, a few seconds of Real Time would equal many years of Mind Time."

Jason was beginning to get that confused look again. So LeRoi changed the subject. "You mentioned if Mind Time speeds up and Real Time appeared to slow down then the converse should also be true."

"What is the converse?"

LeRoi rephrased the statement. "You asked, if Real Time appears to slow down, when the mind speeds up, then does Real Time appear to speed up when the mind slows down?" The answer is obvious. "What happens when you go to sleep?"

Jason thought this was beginning to sound like a Mister Wizard question and answer session. "Well, the mind slows down, and eight hours of sleep seems like only a few seconds." "Except," he thought, "when your dreams beat up on you."

LeRoi caught a bit of the afterthought and decided to try to get him to open-up. "What else were you about to say?"

Jason hesitated. He never told anyone about his dreams. He was afraid of them. He was afraid of what others would think if he ever told the truth. But this guy wasn't just anyone. He was certainly not anyone like he had ever known, anyway. Sometimes when he left the bar he didn't even know if LeRoi existed.

"Well, sometimes when I dream, it's more like a… umm, it's kind of like a..."

"A visit from a dead relative or friend?" LeRoi asked.

Jason was surprised. Maybe this guy can read minds. "Yes," He replied. "Maybe since you're so smart you can tell me how I'm able to talk to my father who has been dead for eight years. Maybe you can tell me how you know what I'm going to say before I say it."

"It's a gift," said LeRoi. "And it's not what I can do but what you can see. The things you're telling me are typical of travelers. People who have the ability to travel through time also have similar experiences. Visitations, at least that's what they're usually called, are fairly common." LeRoi realized he was getting ahead of himself and tried to pull the conversation back to the semblance of a lesson. He looked across the table and realized Jason was back in his imagination.

The kitchen was faded yellow and the double basin porcelain sink was pitted from the many thousands of dishes and pans that had been scrubbed clean by various doers of dishes. Jason's mother stood at the sink, apron around her thin waist and dish soap bubbles up to her elbows.

"When will Dad be up? I want to play catch." said young Jason.

"Your father won't be up until three, son."

Jason didn't know what three was. To him it was just a number and at five he could hardly distinguish one number from another. "How long is three, Mom?"

Always questions; six others before him, but never this many questions. "It'll be soon enough," she answered.

"Why does Dad sleep all day?"

"Because he works all night."

"But all night just takes a blink. I shut my eyes and when I open them it's morning."

"It just seems that way, son. Now go outside and play."

The old conversation had never made sense to him until now.

As the present raced to greet him, LeRoi appeared across the table.

"What is time to a child?" Jason asked.

"Children are special. Until they are taught to read Real Time by clocks, they are on Mind Time. That's why when you ask a person to relate their first memory, they can seldom remember how old they were. They just remember the experience. When you're a child, you explore endless adventures over a never-ending expanse of time. Never-ending that is until you're taught to synchronize your Mind Time clock with the clock on the wall,"

A woman walked by. As Jason watched her move across the room she turned and looked back. She smiled and turned and sat in her booth.

It was already late. The bar was crowded and noisy and he was still on his first beer.

"I need a beer," he stated and away he went. On his way back he stopped at the booth where she was sitting. When he looked back to his booth, he saw it was empty.

LeRoi kicked and let the river take him. "No lesson tonight," he said to no one in particular. Not that he hadn't made progress, but he'd not been able to stick to the topic long enough to begin a new lesson. He knew he couldn't compete with the sex drive of a traveler. Travelers were emotional, easy to anger, easy to laugh, easy to love and easy to hate. They had a roller coaster of emotions which often led to disaster. He knew them all too well. He knew himself.

He wondered if he could teach Jason or if this was all just a waste of time. He knew Jason was living with the wrong person. If he ever found the right one the lessons would be over. So many questions, he thought. It would be much easier if people were predictable, but they never were.

Jason never made it back to his booth. He had been captured by the night. It wasn't planned, at least not consciously. He had been looking for something but never quite knew how to go about it. Being in need all the time had become as commonplace as breathing.

Although everyone bragged at work about their sexual conquests, he had never had any luck. A rookie was what he was. A rookie at about everything, including sex.

It had been exciting...for Jason anyway. A short ride to a cheap motel. A wham-bam, that was swell, had a great time, see you later, type of affair. When it was over, he didn't even realize how it had occurred or whether it was he or she who had been hustled. He thought he should feel guilt or shame or something, but all he felt was a need to get home and get some sleep.

He dropped his clothes in a heap and was quickly asleep. It wasn't an easy night. What should have been a restful night after a week of exhaustion became a jumble of recurring events. His mind seemed to do a typical filing of events but then the nightmare began.

He was standing on the Hill at work. Everything seemed to be running smoothly when a tangle of hot copper rod somehow grabbed his leg and dragged him down into the pit. He struggled but his limbs were leaden and his movements sluggish. He hit the bottom of the greasy pit and the hot copper pulled him slowly back up, towards the waiting stainless steel rolls. He fought to no avail, as ever nearer he was pulled to the waiting giants.

He awoke with a scream. He lay there until the fear started to subside. He reached down and felt his legs, assuring himself he was OK. He was relieved and thankful but afraid to close his eyes again. Afraid of slipping back into the nightmare.

His wife lay on the opposite side of the bed. Awakened by the scream, she wished to be married to someone normal.

Chapter 7

The Monday morning alarm shocked him awake. Somehow, he felt as if he hadn't slept all weekend. He dressed quickly and headed for the copper factory. Both nights he'd had the same dream about being caught in the hot copper. It varied some but the ending was the same waking up in a sweat just before being pulled into the grooved rolls. "Not to worry," he rationalized to himself, "The big black man won't let anything like that happen to me."

John was waiting for him in the cafeteria. It was called a cafeteria but amounted to a long, narrow stall. There were vending machines on each side. Milk, coffee, sandwiches and doughnuts on one side. Soup, candy and microwave on the other. On the far end was a dollar bill changer and on the near end, sat a chest-high space heater. John was leaning on the heater eating a sandwich he'd pulled from a large brown grocery sack.

Some people brought lunch boxes, and some brought small brown paper sacks. John brought grocery bags. Jason put money in the coffee machine, pushed the button and watched the black fluid pour into the paper cup.

"S'pose you can teach me anything today'?"

"Yous ain't gonna be wit me no more. Al says yous going to T-Bone today."

Willis T. Bonnet was his name, but everyone called him T-Bone. Some men went to bulk with heavy work and some went to string. T-Bone had definitely gone to string. He was tall, slim and wiry. He was one of old Al's trainees and probably the best sticker on the Hill. Jason didn't like surprises and it seemed every day was a new surprise. He was just getting the hang of the first station and now they were moving him.

John saw the worry in Jason's eyes and tried to calm him. "Yous got to learn all the stations if yous wants to be a mill man."

"How come you only stick the first station?"

Big John had to think on that for a while. Everyone else rotated through the different stations but John only stuck at number one. "That's cause I's the best at it." The real reason was that no one liked the extra weight on the first station and John had never mastered the flimsier stations.

John continued working on one of the sandwiches he'd pulled from the big brown paper bag. Jason was eyeing the sandwich with a hungry look. As usual, he hadn't taken time to eat breakfast. "You wouldn't have an extra sandwich in there would you?"

John gave him a hard look. "A man's food is like his wallet. Yous don't get in there unless yous asked to!"

Jason was surprised. He took a step backwards. How could a guy go from being friendly to being an asshole in the blink of an eye? He'd obviously made some kind of blunder and stepped over some invisible line, but he hadn't meant to.

The next instant John was back to normal. "The second station is the same as the first as far as timing is concerned. Yous just watch the bar I give you. When the rolls grab it, it jumps. After it jumps you counts to two and catches it. Remember, don't watch my tongs, watch the

copper." It was fair warning, but it probably wouldn't matter. Experience would give him the lesson.

Jason walked to the Hill. T-Bone was there waiting for him. They were the only two at their station. T-Bone reached out to shake hands.

"You be Patch, I reckon."

"Jason Strider's the name."

"Don't much matter what you calls yourself. We're the ones that will be saying it the most." T-Bone looked to the front pulpit and nodded to the dark glass and the old roiling mill came to life. The big bronze couplings rumbled their way up to speed and cooling water came gushing forth from the guides. T-Bone reached down with his tongs and adjusted the splatter skirt that kept the cooling water off his shoes. He reached for the whistle rope, gave a short blast and a long slow blast calling for the first bar of the day.

Jason heard the sound of the furnace door open and heard the big hydraulic ram go into the fire bringing the first bar to the waiting rolls. Big Bertha groaned as she accepted the first bar of the day. He looked around anxiously. He and T-Bone were still the only men on the Hill. There were four stations to be manned today. John's first station and the finish station were still empty.

Jason looked over the top of the mills to the cafeteria and saw John hurrying towards the aisle to the Hill. He had heard the whistle and was trying to move as fast as possible without spilling his coffee Jason listened as each tilt table and each roll brought the first bar ever closer.

Jason looked down the long aisle back to the locker room door and saw the finish sticker come through on the run. He was holding his pants up with one hand and his loose shoelaces flopped as he moved hurriedly towards the Hill. He was trying to run, but with tongs in one hand, and holding his pants up with the other he could achieve no more than a sloppy gallop. Jason could also see the curse words spewing forth from angry lips. He glanced up to the catwalk and noticed Ken Morehead standing at the rail.

Ken glared down from his lofty perch. An opportunity was at hand for someone to get a good ass chewing.

Jason looked to the back of the mill and caught a glimpse of old Al on the back catwalk. The overhead crane had moved up into a position for viewing and Jake, the furnace operator, was standing atop a short stack of bars watching the show.

The first bar had made its way to the long flat entrance table and was racing for the first station John was making his way up the Hill aisle, a few feet from his position. The bar hit the rolls as he was stepping in. He tried to sit his coffee down on the deadman but in his haste he spilled it as the nose of the bar came racing out of the exit hole. John reached, caught and turned. He gave T-Bone a hard look before sticking it.

"Yous owes me a coffee," he yelled over the noise of the rumbling mills. He slammed the hot nose into the waiting rolls.

The finish sticker wasn't so lucky. He was hurrying up the last twenty feet as T-Bone turned with the bar. With a flick of the wrist T-Bone stuck it to the empty station. The nose of copper was out of the hole and heading for the basement as the finish sticker, Billy Joe, reached his position. He was still swearing a steady stream of obscenities as he reached for the errant copper rod.

He reached for the bar, grabbed it thirty feet from the nose and dragged it across the deadman. He tapped it flat with his ax, lifted, cut, double caught, spun and stuck, just as his pants fell to his ankles.

Everyone in the mill was laughing now, everyone, that is, but the hillbilly who was standing at the finish mill with his pants down to his ankles and his boxer shorts flapping in the mill breeze. The scene was a strange contrast to the reality of the job. Jason glanced to the catwalk in time to see Ken turn and go back in his office. Jason couldn't see to the back pulpit, but he was sure Sky was back there shaking with laughter. T-Bone reached back and gave two short blasts of the whistle and the mill rumbled up to full speed.

"You could've waited a few seconds before sticking it"

"Ain't my job to wait on hillbillies. Besides, if you wait too long the loop will hit the basement and the bar cobbles. If the bar cobbles you end up with a pit full of scrap. Whoever ends up with the scrap gets the credit for it."

T-Bore stepped out and handed Jason a pair of tongs. Jason stepped in and waited.

John caught the next bar, turned and made eye contact with Jason, then stuck the bar. Jason saw it jump, counted two and reached into the dirty water for the bar. He was a bit fast and caught it short. He worked his tongs back far enough to stick it and passed it to the finish sticker.

The rod was a lot lighter at the number three station. Jason caught the next one long, and the third one was just right. He was beginning to feel comfortable with the timing but was still stiff and awkward.

T-Bone tried to help him with continuous conversation "Relax and move with the weight. When you bring it up to a point of balance, release pressure on the tongs and slide up past the bend so you can stick it straight. Let the weight turn you and just flow with it. Be thinking about the hole you're going to stick it in before you get there. Let gravity bring you down to the hole and make it all one smooth motion."

The advice was all well and good, but Jason couldn't find the rhythm. He was struggling and clumsy. He was making it but not yet part of it.

It was a long day and the shower after work never felt better. He had only missed a couple of bars and he had managed to cut them both and stick them without forcing a shutdown. He was totally fatigued as he let the hot shower pound away his aches and pains. "How am I ever going to get used to it?" He wondered.

He looked at T-Bone across the steamy shower, laughing and goofing off. He wasn't even tired. Big John was there also. He seemed as fresh as he had been in the morning. Jason felt like he'd been in a fight. Every Muscle ached.

That night the dream came again, worse than ever. This time, in his dream he was being pulled toward the rolls and fighting. He broke loose at the last moment, jumped for the aisle and rolled. Scrambling he got to his feet and ran, dead into an unseen wall.

The jolt woke him, and he looked around. He was laying in the hallway leading to his bedroom. There was a fresh hole in the plasterboard where he'd rammed his head into the wall. There he lay, naked and sweating. His wife stood looking at him. She just shook her head and walked away.

"Not much longer," She thought, "Not much more of this crazy son-of-a-bitch." She headed for the kitchen.

"Not much concern," He thought, "She could have asked, 'Are you all right?' Did you hurt yourself?'" She'd given him that same mean look he'd seen so many times before. He pulled on some clothes and headed for the door. What a way to start the day.

T-Bone was waiting for him on the Hill. John and the hillbilly were in position. T-Bone reached back and blew the whistle, calling for bars. He stepped out and handed Jason the tongs. Jason stepped in and in a joking, cocky manner said, "No problem, I can handle it"

T-Bone turned and walked away. All of a sudden Jason felt very alone, very naked. He looked to the finish station. The hillbilly had a smirk on his face. Jason looked to the first station and John shrugged his shoulders.

Ken was waiting for T-Bone in the cafeteria. "It's
pretty early to be leaving him alone."

"It's gotta happen sometime. He's got all the moves, he just needs to relax and put it all together."

"Don't get too far away. We're forty tons behind this week and we've got two specials we haven't got to yet."

Ken walked away and T-Bone leaned against the old space heater and looked up at the stickers' heads just barely visible above the rolling mills. What could he tell Patch to make him relax? What could he show him which would get him over the hump? Time was the answer. He smiled to himself. Time had been his best teacher. The time it took to make ten-thousand catches. The time to cut a hundred missed bars. All things would become obvious in time. Either Jason would master it, or he'd go away, but it wouldn't take much more time either way.

Jason struggled with every bar. He held on for a long hour with no relief. The lead foreman walked up the aisle and stepped into the finish station to relieve the hillbilly and begin the first round of breaks.

"Just fifteen more minutes without a mistake and I'll get a break," Jason thought to himself. He was already feeling the fatigue. His arms felt weary and heavy from the tenseness.

Billy Joe headed for the cafeteria and bought a coffee. You gonna do anything today T-Bone?"

"I am doing something; I'm breaking in Patch "

"Well, why don't you share the wealth a bit. You spell me at the fourth station, and I'll spell John at number 1, and we'll all get a piece of 'extra break time'

"Don't matter to me," T-Bone replied as they both walked for the hill.

Billy Joe stepped in at the first station, Jason was at number two and T-bone took over at number three. The foreman and John headed for the cafeteria.

"Five more minutes," thought Jason, and "I'm through my first solo without a mistake."

He was starting to feel as if he might make it when Billy Joe turned with the bar. He slammed it into the waiting entrance rolls in a smooth motion. The difference in timing between the big man and the skinny hillbilly was enough to throw Jason off a count, but he made the catch anyway. Billy Joe hesitated on the next one causing Jason to catch it long. He struggled to work his tongs up the flimsy piece of hot copper and stuck it.

T-bone noticed what was going on but there was nothing he could do. Out of character for the hillbilly to want the first station. "Don't watch his tongs, watch the bar when it Jumps," he yelled over the noise of the rumbling mills. T-Bone looked for a way to help. He could see the foreman and John in the cafeteria talking. He looked for old Al or Ken or anyone, but it was too late. They were alone on the Hill.

"Watch the bar." Jason heard the words echo in his mind as he turned for the next bar. I am watching the fucking bar he thought as he got in position to make his next catch.

Billy Joe turned with the next bar. Child's play he thought to himself. "Stick it fast, stick it slow, fake the next stick. Laugh at me for being late to my station," he thought. He laid the nose of the bar in the entrance hole, released pressure on it and slid his tongs hard into the waiting entrance guide.

Jason thought he was watching the bar but was fooled. He saw the tongs slam into the guide, counted two and then reached into the dirty mill water for the bar.

Billy Joe waited. When he saw Jason bend down for the catch, he stuck it for real.

Jason closed his tongs and came up with nothing but air. He looked back to the first station to see if the bar was running. It was. He could see it feeding in. He could also see a smirk on Billy Joe's face.

When he looked back to the exit hole where he should have caught the bar, he saw the nose of the bar was already thirty feet away and heading for the basement. Jason grabbed it, dragged it across the deadman, taped it flat, cut it, released it, slid his tongs to the new nose, turned and stuck it to T-Bone.

Jason turned for the next bar, but it had already been stuck. There stood Billy Joe, arms folded, cradling his tongs as if everything were just fine.

Jason reached for the copper. This one was only twenty feet gone. Reach, drag, tap, cut, stick, and turn. Now there were two pieces of scrap laying in his pit. He quickly turned but he was too late again. It was only ten feet gone. Reach, drag, tap, cut, stick, turn. This time he was in time. He saw the bar jump, counted two and made the catch.

"I did it," he thought, as the three pieces of scrap in his pit began to cause a cobble. One of the bars running in Jason's pit caught one of the pieces of scrap. At first it was a kink, a loop in the bar, like a kink in a garden hose. It hit the entrance guide and tore off. Four pieces of scrap in the already crowded pit. With one of four holes blocked off, Jason didn't know what to do.

T-Bone was watching but this problem was beyond anything he could solve with words. If he'd been with Jason, he could've shown him some tricks to do with a fast axe and a strong arm. It was too late now to do anything but watch.Jason turned and looked to Billy Joe just as he stuck the next bar. The middle station was a mess now. The running bars began to roll into a large cobble.

The second of four holes plugged and tore off. Jason turned to stick the bar but there was no hole open. Two bars running and the remaining two holes plugged. He dropped the bar and reached back to the emergency rope and gave it a blast. He turned back to Billy joe in time to see him stick another bar. He couldn't believe his eyes; He already had a pit full of scrap and here came another bar.

He wanted to catch it, but he hadn't been able to get rid of the last one. He let it go and watched the nose come out of the hole and race for the basement. As he stood there

feeling helpless, he realized the entire giant cobble was moving out of the basement.

Some of it had turned to an ugly black, most was still a fiery red. It looked like a giant web of tangled red hair. He tried to back away and felt the big mills against his back. The bars running to his right were pulling the cobble up towards him. He turned to his left in time to see the hillbilly give him yet another bar. "What the fuck?" he screamed. His eyes grew large as the cobble drew ever closer. He panicked and froze. A few seconds had become an eternity.

T-Bone had been watching every move, every mistake. He had been hoping for a miracle, that's what it would take. An experienced sticker could have cut his way out of trouble and got away with a short delay. A new sticker had gotten buried and T-Bone was left with no options.

He reached for the emergency stop button with his tongs and waited, hoping the bars would tear off and the cobble would stop short of the deadman. He waited until he could wait no more. He had to shut her down or watch this youngblood get fried. He hit the emergency stop.

It was an awful sound. The entire group of mills stopped immediately. Large brass couplings sheared; one took out a water-cooling line. The fountain sprayed on the cobble and created a cloud of steam which rose over the mess like a brown ugly smear.

Ken was already halfway down the stairs. John and the lead foreman were hurrying towards the Hill. Jake was busy cranking down the furnace controls. Sky was attending a large array of blinking red lights. The overhead crane and the crane chaser were both hurrying to the front of the mill.

Each one had heard the awful sound. Each one knew it would be a long hot job getting the old mill up and going again. Someone was for sure in for a good ass-chewing.

Jason let out a long, slow breath. He figured they would all be coming over to see if he were OK. He looked at Billy Joe who had leaned back against the now silent mills and was lighting a cigarette. Billy Joe's pit was empty of all copper. He looked to T-Bone. He was pulling on a pair of leather gloves. T-bone reached down and grabbed a gob of tallow and begin rubbing it into the gloves. His pit was also void of all scrap. Ken was the first one to reach Jason's station and Jason realized it was probably not to check on the condition of his health.

Kens' face was red. The veins in his neck and forehead stuck out like angry blue ribs. The light reflected off his glistening head. He seemed to be coming in slow motion as each step brought him closer to Jason. His first words brought Jason back to regular speed.

"You stupid, fucking, clumsy, idiot. Can't you do anything without fucking it up? It's not good enough for you to cause a delay, you shut the whole fucking mill down."

He went on to explain in spades why Jason would never amount to a hill of shit and how he'd never make a pimple on a good sticker's ass. How, if it would have been him at the finish station, he'd never have touched the button and how the whole world would have been better off if Jason's old man would've kept his pecker in his fucking pants."

Jason had been chewed out before but this one was one for the record books. He stood with his arms folded and listened. The entire group was busily rolling scrap, cutting copper and dismantling mills to fix the broken couplings.

Each one worked ignoring the ass chewing as if they were deaf. But Jason wasn't deaf. He was cataloging every insult. He was filing them all away. Someday he'd give them all back. They all fucked me, he thought. Every fucking son of a whore was against him, he felt, as he listened.

The mill didn't start again that day.

Friday morning finally arrived. Thursday night had been long and restless. His dream beat up on him again. This time it was more real than ever. This time it didn't take much imagination to see the hot tangle of copper moving towards him, he'd seen it for real.

On his drive to work he thought about quitting the mill. It would be so easy to tell Ken he was cancelling his bid. Monday he could be back in the wire mill. His old set of wire drawing machines would be waiting for him. No more bad dreams. No more trying to fit in where there was so much conflict.

John was standing in the cafeteria eating a sandwich out of his large feedbag. 'You's be starting on two today. The hillbilly ain't coming in." John could see the worry, the fatigue. "You's just got to learn to relax. You's making it a lot harder than it is."

"I've been thinking I should turn it down." Jason said while looking down at his shoes. "I could be back in the wire mill Monday. We might all be better off. I keep having bad dreams. I can't sleep, I can't relax, and I can't seem to do anything right."

John wasn't surprised. He'd seen them come and go. Some made it, some didn't. Once they started whining, once they sat down on the pity poddy, they didn't last long.

"Maybe I should get me a violin and plays you's a tune while you's sing the blues."

Jason's temperament went from self-pity to anger. He wanted to strike out at someone. He wanted to tell them all how fucked up they really were. John saw the change. Anger he liked. Self-pity made him sick. "I've put some time into you, boy. Every bar you missed made me work harder. Every mistake you made, made me sweat a little more. If you want to leave, go ahead, that just means I'll have to break in another. As for the dreams, you think yous the only one that's had em? You think yous the only ones what's ever been scared."

"I ain't scared"

"Yous a liar. Your dreams call you a liar. Old Al says you be a good one if you don't quit, but it don't matter if you quit or not. Today you'll be on number two cause yous all we got. Monday will get here soon enough, whether yous here or not. Today's what yous got to take care of."

John went on eating his sandwich.

Jason angrily headed for the Hill. He didn't like being called a liar, especially when it was the truth. Ken intercepted him. "You'll be on number two today. T-Bone will stick finish and John will be on number one. Don't fuck it up."

T-Bone was waiting for him. "There's two pair of second stand tongs on top of the mill. If you lose a pair, just grab another and keep going. I'll be quick on the whistle today. If you miss two in a row, I'll blow it. We can get away with a shut down but if we have to go to emergency stop again, Ken will have all our asses up front."

John was on his way up to the first station as T-Bone blew a short and a long, calling for the first bar.

Jason reached down with his tongs and adjusted the splatter skirt. He shifted his weight from one foot to the other waiting nervously for the first bar. After yesterday, he was afraid to even reach for it. He was afraid of failing.

It came. From the time he heard the furnace door open, until he saw it racing down the table to John's station, it came. Wishing it wouldn't get there didn't slow it down. In fact, it seemed to speed up.

It hit the first station with a thud. John grabbed it, turned, made eye contact, and stuck it.

Jason went down. He reached into the water and came up with a perfect catch. It seemed so light, so easy.

He released pressure while he turned, slid his tongs up past the bend, caught sight of the entrance hole before he got there, and made a nice smooth one stroke delivery. He turned for the next bar and realized he couldn't even see it yet.

T-Bone reached back and blew two shorts.

Jason couldn't believe how easy it was. His mind had remembered all the misery and mistakes from the day before.

The present, today, didn't seem to have any resemblance to the past. He caught every bar. It became like a game of catch. It was so simple. After an hour, the lead foreman came up the aisle and relieved him. Jason almost wished he didn't have to go on break. It was as if he was afraid to take a break for fear of forgetting.

He headed for the cafeteria and bought a vending machine can of soup. The assistant foreman relieved John and he met Jason in the cafeteria.

"I got it now, John. I got the rhythm. It's easy."

John busied himself with his second sandwich of the day. "Nothing to it' Patch. I sticks it and you catches it. Yous just keep your concentration and everything will go easy."

He wanted to caution Jason about becoming too relaxed, but he knew it wouldn't help. Learning would come one lesson at a time. It had always been that way. Just when he thought he understood something the old evil enchanter would give him a heaping helping of humility. "Yous just keeps on your toes." He said as he chomped into a big ham sandwich.

Jason headed for the Hill. He relieved the lead foreman and the lead foreman in turn relieved T-Bone.

T-Bone met John at the cafeteria. "Kid looks pretty good today. How long before he digs another hole for himself?"

"I don't know, T-Bone, how long did it take you?

T-Bone couldn't remember that far bark. He'd made so many mistakes and cut so many bars. He just shook his head.

"Well, we're doing all we can. I told both foremen to stay close. I don't' want another mess like yesterday. I should have known better than to let Billy Joe ahead of him. He was really mad at me for starting early yesterday."

"Don't matter none. It had to happen sooner or later. If Patch can't learn to handle the heat he might as well go back to the wire mill."

"Did you see him while Ken chewed him out? He never said a word. Ken really out-did himself on that one"

"Yea, I saw him. He didn't even try to lay it off on Billy Joe, not that it would have done him any good. Scrap belongs to whoever ends up with it. Patch had it all. Still...he had a look about him, like he was saving something for Billy Joe, and maybe the rest of us. What you think he's up to?"

"I don't know, John, but you know he can't come up with anything we haven't seen before."

John finished his sandwich and headed for the Hill. Jason was working like a well-oiled machine. He was beginning to develop some style. He was also beginning to relax.

Every eight seconds watch the bar jump, reach, grab, double catch, rotate back down and stick. "Piece of cake," he thought.

John was at number one, Jason at number two and T-bone was back from break at number three. Ken was relaxed in his office smoking a cigarette. Sky was jacked back in the rear pulpit drinking coffee and chewing apple plug tobacco. Old Al was out on the bar dock, sitting with his back against the old brick building, reliving some piece of his past.

All seemed right with Jason. He was being lulled into complacency the way a driver gets lulled by the short dash lines dividing the lanes of a highway. He went down, made his catch came up with the bar, released pressure on the tongs and slid them up past the bend to complete the double catch. As he tightened on the bar, he realized the bar was gone.

The bar had floated momentarily. This was only a very slight mistake; just enough for the hot copper to escape the jaws. Gravity was in control now. The free-floating red-hot nose dropped onto Jason's wrist...SSST. Kiss number one. The realization of not having a hold on the bar came almost simultaneously with a hissing sound and the acrid smell of burned flesh. The hot pain shot up his arm from his wrist as he let go of everything. He was awake now. Awake and aware. Suddenly real time seemed to slow down. His tongs clattered to the steel deck and began a slow slide down the greasy pit along with the nose of the bar.

Jason reached back for the spare tongs. He bent down and grabbed the hot bar thirty feet from the nose. He dragged it back across the deadman, tapped, cut, re-grabbed, spun and stuck. He spun for the next bar and John was holding it for him. John's eyes were saying: "Slow down. Relax."

John stuck, Jason caught, turned and stuck. He was angry. He Looked at his wrist and saw the burned cracked skin. He could see pink beneath the charred flesh.

It had happened so fast. In trouble for an instant and then he'd cut his way clear. He was out of a jam with only a black spot on his wrist. He looked at the thirty-foot long piece of hot scrap in his pit. He glanced back to John and noticed he was there with another bar. Jason caught it and stuck it. He reached for the scrap and with an angry snap he tried to clear it to the aisle. He snapped it much too hard. Instead of floating up and away and out of the steel slide, the bar that had already kissed his wrist came flying at his face.

He shied away from the blackening tip of the bar, but not fast enough. SSST. Kiss two. The piece of scrap slapped him across the ear. As he dodged to get away from the bar, he banged his head against the mill cap.

It was like a one-two punch. This one didn't hurt a bit. He felt lightheaded, like he was floating. He felt his knees go loose.

T-Bone was watching carefully. He saw Jason's knees start to go and reached for the emergency stop. If Patch fell into the fast running bars he'd have to shut down or watch this youngblood get cut in half by the hot copper.

Jason's eyes went blurry. He reached out with his tongs and found the deadman. Using them like a crutch he held his balance. From somewhere a long way off he heard his name.

John was standing there with a bar. "Patch," he yelled.

Jason looked over at John. He could see two big black men. He shook his head to clear it.

"Watch the bar," John yelled, as he passed it to Jason.

Jason went down, down to get the bar... To his surprise, he came up with it. He came halfway around, stopped at the deadman to straighten it and then turned and passed it to T-Bone. His coordination was off, and he took too long a stride. His right leg pushed into the fast running bar at the inside position. Just a fraction of a second and the bar cut and burned its way through Jason's pants leg and through two layers of tight skin. SSST, Kiss three.

He was dancing on one foot, but the pain cleared his mind as he reached for the next bar. One legged, he caught it and stuck it. It hurt so much he wanted to cry, but he didn't. He was mad at himself for the lapse in concentration. Jason slapped at the piece of scrap that had kissed him on the ear. He knocked it the rest of the way into the aisle. He had no one to blame but himself. He looked around to see if anyone was watching.

Not only were they all watching, he realized they were all laughing. T-Bone was holding his tongs like a banjo. He was pretending to play a tune. The other mill men were clapping like there was an old-fashioned hoe-down occurring.

"You dance, Patch, and I'll play the tune," T-Bone hollered over the noise.

Jason quit dancing. At first, he was furious but then as the pain started to subside and the fury faded, he almost smiled. Ain't so bad, he thought. Ain't so fucking bad.

He had been kissed and survived, the pain had been identified and defined. It was now just a positive incentive for not making mistakes. He somehow felt he was more a part of the group. It felt good to belong, if only a little.

Maybe this ain't so bad back here. Maybe he could get along. Maybe he'd just keep quiet about transferring back to the wire mill. He wouldn't look at his leg or rub it. Somehow it was like being hit by a fastball. If you rubbed it, it was like admitting a weakness. Didn't really hurt, he told himself. The lie made him feel better. That night the dreams went away.

Chapter 8

LeRoi pulled his canoe on shore and began the short climb to the bar. No need to secure the boat. In this day of cars and motorcycles, no one ever noticed or realized there was such a thing as a river traveler. The Old Mill bar had once been a grist mill. It provided a service for the local farmers in the early eighteen hundred. Those were sometimes. Indians and canoes. Farming and scratching a living out of the land. The environment was clean back then and the people dirty. Now it was the other way around.

He tried not to let himself regress into all the modern problems of the day, pollution, crooked politicians, corrupt law enforcement and most of all, waste.

The problems of the twentieth century overwhelmed him and he often longed for the other life. How long had he been back? Just eighteen months, it seemed a lifetime ago. Reestablishing himself within this structure called the American culture had not been easy. Two years ago, he had given up on America. He had looked at the waste and decided he could no longer be a part of his own country. He decided to go to the mountains and become an island unto himself.

A hermit is what he wanted to be. He wanted to live his life without being a destroyer. He sought to consume no more than necessary to survive. He'd gone to the mountains to be himself. To escape the injustice and stupidity.

The words of Thoreau were his guide. They had jumped out of the old book and he remembered them well. Henry had defined the existence and LeRoi had tried to copy it. LeRoi remembered the quote word for word:

"I went to the woods because I wished to live deliberately, to front only the essential facts of life, and see if I could not learn what it had to teach, and not, when I came to die, discover that I had not lived. I did not wish to live what was not life, living is so dear; nor did I wish to practice resignation, unless it was quite necessary. I wanted to live deep and suck out all the marrow of life, to live so Spartan-like as to put to rout all that was not life, to cut a broad swath and shave close, to drive life into a corner, and reduce it to its lowest terms, and if it proved to be mean, why then to get the whole genuine meanness of it."

LeRoi smiled to himself thinking how green he had been. Just put a bedroll under your arm and stick out your thumb and hitch for the mountains of Tennessee. He had a plan. It had seemed a good plan at the time.

His solution had been to escape and experience a better way. He'd stuck out his thumb and hitched to the foothills of the Great Smoky Mountains.

To be a combination mountain man and hermit were honorable goals. To achieve those goals had been an adventure of learning.

What he'd learned first was how miserable it could be in the mountains when it rained and your only tent was a garbage bag.

He'd learned how tough it was to catch small game and how dumb he was about what was edible. He quickly learned what would sour his stomach and cramp him up for a very long night.

The elements drove him down from the mountains and into a library in Gatlinburg, A few weeks of study and some sparse supplies and he was back in the hills. It was easy after a while, to set a camp and learn a territory... to live off the land and meditate and travel within the regions of his mind. At night he would explore stars in a moonless sky and travel forever while lying on his back on a hillside, knowing no bonds. His isolation became addictive and the sight of another person would make him uncomfortable. The occasional hiker who would happen by close to one of his territories would never see him. Like a skittish deer he would observe from a distance, never giving up his position.

He often felt like a child playing hide and seek with the world. His only contact with people occurred during his journeys to the past. The itemization of each of the events in his past were rehashed and chronicled. Each conversation was re-talked, each experience relived. It happened slowly. One adventure at a time, and unknowingly, he was training himself to travel in the past.

The more he trained, the better he became at time travel. One morning he awoke to the real world and realized he had barely the strength to crawl. His journeys to the past had stolen so much of his present that real life was but a bare and fragile thread.

Suddenly he wanted to converse with another human being, not a figment of his past but a real, live person in the present. He had journeyed and explored the mysteries of the past and now he had a revelation as to his goal in life. He realized if people understood the principals of time travel, they might gain some insight into their existence. If people began to see the patterns of life as he had, they may begin to live. If he could explain the basics to the world, there might be hope. Hope is what brought him back.

He remembered crawling to a nearby stream for water and when he saw his face in the quiet pool, he was shocked. He knew he was close to death. Two sunken eyes within a mass of hair. How long had he been time traveling? He didn't know. He had lost track.

The water refreshed him and gave him strength. Not far away was a berry bush and that saved his life. His choice for life was all he needed, and he slowly but surely made his way back to Gatlinburg. He had gone to the mountains a young man and within six months he had become ancient. At a soup kitchen in town he gained nourishment and within a month his body recovered to its youthful self. His eyes though, would carry a strange gauntness forever.

What a fool he'd been to risk his life out of bitterness… just go away and die for lack of interest... but, even a fool gains wisdom if he's allowed to continue in his folly.

His years at the university had given him enough of the technical skills of science to last him a lifetime. His six months in the mountains had given him a hope for the future.

LeRoi had more of a vision than a hope maybe, but whatever it was, it saved his life. Now he had an obligation to pass on the information. The problem was, who do you pass it to? Who cares enough to want to be a part of this fuzzy outline for the future?

He had come to the conclusion that he could only pass the information to a fellow traveler and finding one would be a problem. Then he'd met Jason.

He still had no clear idea how to get to where he wanted to go with his lessons, but he was communicating to someone.

To communicate was a blessing. To find one who speaks your language was like finding an oasis in the desert. He climbed the hill leading to the bar. Maybe we'll get in another lesson tonight, he thought.

Chapter 9

Frank looked out the dingy window to the parking lot and saw LeRoi walking to the bar. There were no cars in the parking lot except for his old Mercedes. There's Jason's friend, he thought to himself. He reached for the bottle of Ouzo and a shot glass.

LeRoi walked up to the bar and slid a dollar to Frank, watching as the clear liquid filled the shot glass.

"How's it going tonight, Frank?"

"Pretty slow right now, but business always picks up by eight. What are you up to this fine evening?"

"Just here to give Jason another lesson in time travel. Later on, we'll solve most of the problems of the free world and then we'll move on to more serious things."

Frank laughed and busied himself with the regular routine of filling coolers, restocking liquor and setting up glasses.

LeRoi took a small sip. "Heard you were on the outs with your wife. How do you like the inner workings of the justice system?"

Frank stopped working for a minute, as if contemplating some great mystery. "No," he said as he continued working, "We reconciled our differences. Now I'm just like Gene Autry, back in the saddle again. Actually, I talked it over with my lawyer and found that if I went through with the divorce, I'd lose the bar. I decided to fall in love all over again.

"Another marriage made in heaven," LeRoi thought as he picked up his shot and headed for the seclusion of the corner booth at the far side of the partition.

Jason had finished his shift, received his paycheck, been home, made himself a sandwich, caught a raft of shit from his wife, escaped, and headed for the Old Mill Tavern.

He pulled into the parking lot. No one in the lot yet but me and Frank, he thought to himself. As he got out of the car, he knew LeRoi would be there.

Frank had the beer out of the cooler before Jason walked through the door. Jason picked it up and headed for the time machine.

A time machine was what it was beginning to feel like to Jason. He seldom remembered the minutes or hours he spent there but he always remembered the experience.

Jason sat down. LeRoi noticed the brown charred spot on his wrist, the lump on the right side of his head and the burn line across his left ear. There was a band of inflammation radiating away from the burn.

"It happened at work today." Jason answered the question before it was asked.

"You learn fast. You're answering my questions before I ask them. Have you been practicing mind reading?"

"I didn't read your mind, I read your glance." Jason was surprised with the answer, and he followed it with a question. "How much can you read from a glance?"

"Sometimes you can pick up a whole lifetime. Haven't you heard the expression: 'You know your friends better in the first instant you meet, than you know your acquaintances in a lifetime?' It seems to have something to do with the compatibility of the two people rather than the frequency or amount of time they spend together."

Jason thought back to the moment he'd first met LeRoi. LeRoi was thinking of the same moment and neither spoke. The quiet moment was not uncomfortable. LeRoi remembered one of the old definitions of friendship.

"Friendship is when you can be quiet around someone and not feel uncomfortable," he said without hesitation. He didn't want to get off on a tangent, so he jumped to the lesson. "Do you remember the first two lessons?"

Jason had to think back. "Sure, lesson one you said time began, is now and is continuing. Each minute is a minute unto itself and cannot happen again. Lesson two you stated there were two types of time, Real Time, which is measured by the clock on the wall and Mind Time which is a variable. If you speed up Mind Time, Real Time appears to slow down. Oh yeah, and if you slow down Mind Time, like when you go to sleep, Real Time appears to speed up." Jason stopped and took a deep breath. He had surprised himself as to how easy it now seemed.

LeRoi smiled inwardly. This was further than he had ever been able to get in explaining the simple concept. "That's great Jason, tonight we're going to learn about the past.

"If each minute is a minute unto itself and if you can't go back and change it, what good is learning about the past?"

"I didn't say you couldn't go back. I just said you couldn't change it."

"That's not a whole truth," he thought, but he didn't want to confuse the issue right now. "There are many good reasons for going back. You may want to re-assess a situation or change your mind about a conclusion you came to. Your travels to the past can keep you from making the same mistakes over and over again, you may want to revisit a specials place which brought pleasure or pain, whichever happens to be appropriate."

"Okay so let's say I want to go to the past. How do I do it?"

"There is more than one way to travel. That's what we are going to do tonight."

The barmaid came by and picked up Jason's bottle, tipped it check its contents, and asked if they wanted another.

"Sure thing," said Jason. "Bring us both a beer and a shot of Ouzo. If I'm going on another journey, I may as well take some Ouzo with me."

People were starting to filter in, and the background hum was increasing. She brought the traveling medicine and exchanged it for some money. Jason settled back against the wall and sipped the licorice-tasting shot.

As LeRoi spoke, the outside noise began to decrease in Jason's mind. The external influences were fading to zero.

"First, we will talk about physical time travel. It is the easiest but also the most boring. Is there any particular time period you've ever wanted to live; any special time you think you may have enjoyed more than the present?"

"My great-great-great-great-grandfather, Daniel Strider went on the gold rush in eighteen-fifty. I have always wondered what it would be like to live on the farm where he lived prior to going. He was twenty-two years old, had been married a year and had a baby girl. He and his friends formed a company and went west to look for gold."

"Did he take his wife?"

"No. Just he and his friends. He was gone a couple of years and returned to southern Illinois."

"Did he find gold?"

"I'm not sure. If he did, he didn't find much because he came back and lived out his life as an innkeeper."

"I've often wondered what it would be like to work the soil and build a homestead. No electricity, no cars, no pollution or overcrowding." Jason thought for a moment. "No factories, no factory work."

"No problem," LeRoi interrupted. "Tonight, we could physically go there."

Jason hadn't drunk enough to be believing this. Sure, he'd had some experiences in time travel but not physically being transported to another place. "Like, beam me up Scotty, and beam me back down in the spring of eighteen-fifty?"

No, you don't need a tele-transporter or any fancy equipment. We just need to go out, get in your car and drive to Grable."

Grabill was a small town just north of Fort Wayne. There was an Amish community nearby. Jason knew of it. When driving around the vicinity you had to really watch your speed. A careless night of drinking and driving in that area might get you a windshield with Amish horse and buggy splattered across it.

"You could get a job working for one of the large Amish farms. After a few years they might take you into their religion. You could sell all your worldly possessions and you'd be there. Walla, eighteen fifty all over again. No electricity, no automobiles and no factories."

Jason thought about it. He'd been around horses at his uncle's farm. He imagined himself getting up with the morning sun, pulling on his hand-made bib overalls and pumping up an old gas lamp. He struck a match and opened the valve. When he heard the hiss, he put the flame to the mantle and jerked back his hand as both mantles flared. He walked to the barn and hung the lamp on an old nail. He lifted the harness and bridle from their place on the long row of wooden hooks and moved toward the stalls. The smell of leather and horses filled his senses. The huge draft animals shied away from Jason as he readied their burdens.

Their breath was evident in the cool dawn air and they stomped and neighed as he fit them to the reins. He gathered the bridle and harnesses and positioned the animals for the morning work. He gave them a quick walk around before hitching them to the two-bottom plow.

As the sun began to break on the horizon he was in the field. He enjoyed the feel of fresh-turned earth beneath his feet as the six-horse team leaned to their task. The first hundred yards were the toughest as they fought the load, farting and shitting as they settled into a harmonious routine.

Piece of cake, Jason thought as he felt the first blush of sun on his face. Seek your century and chose your fun. Suddenly the farm sounds were buried beneath the sound of four F-15s racing low across the morning sky. Each peeled out of formation and turned south back toward the Air Guard base at Fort Wayne.

Jason looked across the table at LeRoi. "Won't work," he said. "Every time an airplane went over, you'd realize what century you were in."

"Don't you remember the cliché? 'A man hears what he wants to hear and disregards the rest.' Do you think the Amish see or hear those planes?"

Jason had to think about it for a while. "No, the adults probably don't hear them, but I'll bet the kids do."

LeRoi continued. "You can go back further, you know. You can go to villages in Mexico or any of the villages in the third world countries. Conditions there are almost identical to bronze age cultures."

"No good, because you would still have your memory, and you would know about the rest of the world even if no one else did."

"You could go by plane to Africa and bail out near the tribe of the Tshidi. No iron, no agriculture, limited language and stone knives. Keep your mouth shut and within a few years you would be accepted. You could probably even take a mate and live to a ripe old age."

Jason didn't care much for this type of time travel. This was just a way to play tricks on yourself. "Same problem as before. You'd still know about the rest of the world."

"If you really wanted to go back, you could go all the way to the territory of the mountain gorillas. One woman did. It took a while, but after a few months she was accepted into the group. Now that's really going back."

"That's not going back. You'd need to lie to yourself. You'd need to fool yourself every day. You'd have to be blind to your past experiences and to everything you've ever learned. I don't think I could fool myself that way."

"Sorry. If you want to take your body along, then that's the only way to go to the past."

"Your body is tied to the biological clock within, which is tied

to real time. Your body doesn't care where it spends its time, it only cares about the time span. Most of our bodies have about fifty to eighty years. After that, the biological clock runs out and that's it."

Jason was becoming frustrated. This type of time travel didn't interest him in the least. The idea of living among the Amish had been interesting but he knew his wife would never go for it. She loved her bright clothes and fancy car. She loved the night life and action. If his wife wouldn't go, then his son couldn't go.

It was a package deal. His boy was part of the package. Without Brian, Jason didn't want to go anywhere. He quickly disregarded the option. It was a pleasant dream he thought to himself.

"So, what's your point? What's this got to do with the trip I took to my old playground? In fact, I went past there last week. It doesn't even exist anymore. They built a new school where the old playground used to be."

"Sure, it still exists. You went there a few weeks ago when you were sitting here in the bar."

Had it only been a few weeks? Jason had to think about all that had happened. "If it doesn't exist in my present and if I can't take my body back there, then," he paused for an instant, "Well, if there are only two types of time, Real Time and Mind Time...the only way to get to my past is through Mind Time."

"Bingo"

Jason had come full circle again. Just when he'd allowed himself to think time travel might be possible, he'd realized what LeRoi was talking about was daydreaming.

"What do you mean 'Bingo' that's daydreaming. I'm sure as hell one of the best there is at that. That skill and a dollar, and I can maybe get myself a fucking cup of coffee."

LeRoi saw the disappointment leap into his eyes. "If you only knew. If you stay with me, you'll see how powerful daydreaming can be."

"I'm with you. But if you're going to teach me how to be a dreamer, I already know how."

"You have traveled to the past, and so have many others. What no one realizes is the same mechanism which allows you to travel to the past will also allow you to travel to the future. Once you learn to control it, the earthly chains of Real Time that imprison your body will no longer be able to restrain you." LeRoi wanted to tell him more, but he knew better. He wanted to reveal how powerful the tool really was. He wanted to tell him about the ones who had discovered these mysteries, Daniel, Isaiah, and a score of others. He wanted to tell him there was an easy way to change the past, but he knew he had to wait until the proper time.

Jason got up and headed for the restroom. The bar was noisy and crowded. A haze of bluish cigarette smoke swirled beneath the light above the pool table. Jason came back to the table and turned to sit down. He had done a quick scan of the bar in his ever-persistent lookout. Over in the corner booth sat a dark-haired woman. Strange, he thought, women seldom sat in here by themselves. He squinted through the smoke. She looked up and caught his glance and quickly turned away. Not soon enough.

LeRoi followed Jason's gaze. Black loosely curled hair. Slightly built from what he could see. Too much class, he thought, for a bar like this. He looked back to Jason, to the woman. Shit...end of lesson.

Jason walked around the pool table to the corner booth. She was busily working on the keyboard of a new calculator. The warranty card lay next to the packing, which lay next to the box it came in. She didn't look up or acknowledge his presence.

"Like to shoot a game of pool?"

She looked up. "If you continue to bother me, I'll call my brother over from the other side of the bar."

Jason glanced around to the various groups of men. He glanced back to the woman. He turned and walked back to his booth.

LeRoi saw the disappointment in his eyes. "How'd you do?"

"Great, she said if I kept bothering her, she'd sic her brother on me."

"Which one is her brother?"

"Don't know."

"What's she doing?"

"She's fiddling with some kind of calculator."

LeRoi thought a moment.

He knew he shouldn't interfere. Had he seen something between them, or had he just imagined it?

"You want to meet her?'

"That's all I need. First, I get my ass kicked at work and to finish off my day, you want me to get my ass kicked over some woman I haven't even met?"

"You give up too easily. If she really wanted, you to go away she would have told you that her boyfriend was across the room."

"OK Casanova Varita, so how do I meet her?"

"She has a calculator, right? Tell you can add faster than her calculator."

"That's a good lead in, but the problem is I can't add faster than her calculator. Nobody can add faster than a calculator."

LeRoi pulled a calculator out of his top pocket and laid it in front of Jason. He took a pen from his shirt and pulled a piece of scrap paper from another pocket.

"You use my calculator and I'll write the numbers down and add them long hand. You pretend to be her, and I'll pretend to be you. Let's do five, six-digit numbers. You give me the first number."

"This is bullshit. No one can add faster than a calculator...unless maybe you're one of those...uh...you know...One of those nuts that are good with numbers but can't tie their shoes."

"Idiot Savants?"

"Yea, whatever. You're not one of those are you?"

"No, I can tie my shoes. Bet you a drink I can add faster long hand than you can with my calculator."

This will be an easy beer, Jason thought. He opened the front of the calculator and ran his fingers through a quick addition exercise. He cleared the calculator and readied himself for the competition.

"Ok, here's the first number in the column, six, twenty-three, four, fifty-six. He keyed in each digit as he spoke it.

LeRoi shielded the numbers from Jason's view with his left hand as he wrote with his right hand. He seemed to be taking longer than he should to write just six digits. Jason waited until he was finished writing. "You ready?"

"I'm ready. Give me another one,"

Jason took another six digits off the top of his head. "Fifteen, eighty-six, twelve."

Again, Jason keyed in the digits as he spoke and LeRoi wrote them down. This time it didn't take as long.

"You ready?"

"No. I get to choose the next one. You ready?"

Jason began to smell a rat, but they hadn't stipulated who would give the numbers. For lack of a good argument to the contrary he answered. "Sure, I'm ready, shoot."

"Eight, four, one, three, eight, seven. You give the next one.

"Fifty-nine, thirty-one, twenty-two." Jason again keyed in each digit as he spoke. He waited for LeRoi to finish.

"Ok Jason, I get to choose the last one. Are you ready?"

"Ready"

"Four, zero, six, eight, seven, seven."

As soon as LeRoi finished speaking the last digit, Jason hit the equal button on the calculator. The answer appeared instantly on the LED display. Fast, he thought to himself. Faster than fast. Jason had always prided himself with the skill he had when it came to numbers.

Calculators made it so easy. If you could punch in numbers on a keyboard you could be the fastest adder in the world.

As soon as he saw the answer, he looked across the table to LeRoi. He expected to see him in a frenzy trying to add all the numbers longhand. LeRoi was just sitting there. He hadn't moved since the last digit was spoken.

"Well?"

"Well what?"

"Aren't you even going to try to add them up?"

LeRoi lifted his hand off the paper, revealing the answer, "Two million, six hundred twenty-three thousand, four hundred fifty-five. I win."

Jason couldn't believe his eyes. He looked to the calculator and then to the paper, then to the calculator. It checked. "Fuck."

"Ouzo"

"What?"

"You owe me an Ouzo."

Jason motioned to the waitress, "Two beers and two Ouzos."

Jason was confused. He had always liked a good riddle, but he never liked to be tricked. Was this a trick or was he that fast? No. It wasn't fast. LeRoi had written the answer before the last number was given. How could that be? Had LeRoi somehow raced forward in time, gotten the answer, written it down, traveled back and then, continued with the problem?

Bullshit. That's not possible. No one can go to the future and then buzz back like that. He was beginning to get angry. It always pissed him off. Not being able to figure out the obvious, and these types of riddles were usually obvious. "Don't get mad," he cautioned himself. "Think. Go over what happened. Look for clues. What was unusual? What was out of line?"

The waitress brought the drinks. Jason paid from the pile of money laying on the table in front of him.

He looked across at LeRoi. Neither had said a word. LeRoi was also watching Jason. The kid's good, he thought to himself. But he'll not be good enough to figure this one out.

Jason was going over the game in his mind, "I gave the first number. He wrote it down as I gave it to him. He took a long time to write that first number down...That's it! He must have written the answer as soon as I gave the first number. But how? I gave the first and second numbers. He gave the third. I gave the fourth and he gave the fifth. He somehow knew the answer and forced the result with his own numbers."

Click, click, and click. Like tumblers falling into place, Jason's mind did a match with something from a long time ago.

He was back in high school. Algebra, fifth period, sophomore year. His teacher was standing at the board. There were only five minutes left in the period and the lesson for the day was finished. The teacher, Mr. Garrett, turned from the blackboard and asked, "Do any of you know how to add?"

"What a stupid question," Jason thought. He was, after all, a sophomore. He knew all there was to know in the world. He and his classmates were all good students, and this boring old fucker was asking if they knew how to add. The question was so stupid no one had even bothered to raise a hand. The class had, Jason remembered, given each other some knowing glances. The glance said this guy's finally gone off the deep end.

"Crystal, give me a six-digit number."

She did and he wrote it on the board.

"Jason, give me another six-digit number." The old man had written it and inserted the third number. He asked someone else for the fourth number and wrote the fifth number himself. He drew the line under the addition problem and wrote the answer all in one motion. Jason's chin had almost hit his desk. The bell rang. As everyone got up to leave, he remembered old man Garrett talking over the noise. "Anyone wanting to learn how to add like this please stay after class." Jason had wanted to stay, but only snitches and Teacher's pets stayed after class. He was much too hip to take a chance on that type of reputation. He had gone with the rest of the class, no smarter for the experience.

He blinked his eyes and LeRoi appeared across the table. He had seen Jason's eyes leave and was patiently waiting for him to return.

Jason was back but he was silent. How many lessons have I missed because I was too cool to learn? The old riddle had laid there, somewhere in the back of his mind, a little black spot of stupidity. How many other black spots lay there in his mind? He didn't know. He only knew he wanted desperately to start turning on some lights. This guy LeRoi seemed to have control of the switches.

Jason had forgotten all about the woman. After going on his little time adventures to the past he often forgot what he had been doing before he left.

"I've seen the trick before, back in high school." Jason was looking intensely at the column of numbers on LeRoi's piece of scratch paper. "So, what's the trick?"

"You mean you don't believe in magic?"

"No."

"Good. Neither do I. There are much stranger things in real life than in magic. Look at the first number you gave me and compare it to the answer."

Jason saw the similarities immediately. The answer was the same as the first number he had given, except there was a two in front of the original number and the last digit six, had been reduced to a four.

"It's always the same. Whatever number is given first, just put a two in front and subtract two from the last digit and you have the answer. You can write the answer whenever you want. I had to write my answer right away, because I had to have it down before you hit the equal's button."

Jason still didn't understand. "I gave you the first, second and fourth numbers. You gave me the third and fifth numbers. How did you know what numbers to put in?"

"That's the easy part if you know how to add. Look, when I gave you the third number, I made each of your digits plus each of my digits equal to nine. Look at the paper. "Your first digit of the second number was one. My first digit of the third number was eight. One plus eight equals nine. Your second digit of the second number was five, my second digit of the third number was five. Five plus four equals nine. You do the same thing with the fourth and fifth row of numbers. "Simple huh?" A little spot in Jason's mind suddenly lit up. He didn't quite understand the why of it, but he certainly understood the how. He slid the calculator over to LeRoi and pulled the scratch paper to himself. "Let me try it."

Jason asked LeRoi for the first two numbers, inserted the third, asked for the fourth and inserted the fifth. It worked beautifully. Jason turned and looked across the room to the corner booth. She was still there.

LeRoi followed his gaze. "There's only one rule. The last digit of the first number she gives you must be greater than two."

"What if it's not greater than two?"

"Just ask her to keep adding digits until she gives a digit greater than two."

"You mean this works with more than just six-digit numbers?"

"Sure, you can run the numbers clear across the page if you like. Your column can only be five numbers high for this trick."

Jason downed the remainder of his Ouzo, finished his beer for added courage and walked to the corner booth. She was filling out the warranty card. She looked up, saw who it was and continued to fill out the card. She glanced up again and looked across the room to a group of noisy men still, dressed in their softball uniforms.

"Bet you a drink I can add faster long hand than you can add with that calculator."

She looked at him with some interest. "I don't believe that's possible."

"I can't do it standing up," he said.

"So have a seat," she said, as she pulled out a piece of paper.

As Jason slid into the booth, he looked across the pool table to his time machine. The waitress was at the empty booth, picking up the empty bottles and wiping off the table.

LeRoi was already halfway down the old trail. "Why can't you learn to keep your big mouth shut?" He thought. He stepped into the sleek boat and gave a silent push into the slowly moving current. He sat on one cushion and leaned back against a spare cushion he had propped against the center cross brace. The river took him slowly away. Away to some moonless starry night in the Great Smokey mountains.

Back at the bar, the trick had worked great. The dark-haired woman was as amazed as Jason had been.

"How did you do that?"

"It's easy." Jason moved from his side of the booth to her side. "You just concentrate on each number as they're given, race forward in time, get the answer and race back and write the answer down." It was the best lie he had told in a long time. He could tell she was amused. Her defenses were down.

She looked at him as if he were some kind of intellectual. He liked it. To have a woman look at him as if he were worth something. It was a special feeling.

She placed her hand on his leg as she continued to talk to him. The sensation was unbelievable. It was as if her fingers were on fire. He felt a warmth through his blue jeans he'd never felt before. She withdrew her hand, but the warmth stayed. It was like removing an iron from freshly pressed pants and then pulling them on. The heat lingered after the iron was removed. The feeling made him blush. He wondered if she felt it too. He thought he saw a tinge of

color high on her cheeks. Not the color that comes out of a make-up kit, but something fresh. Something from within. Something very deep. Neither one spoke for a moment. He was groping for a sentence.

"Let's get out of here. Can I give you a ride somewhere?"

She thought about it. This was definitely not the kind of complication she wanted or needed. She was scheduled to fly to South America in the morning to be with her fiancé. Another six months and she would be married.

She gave him a long look. He looked like a bar brawler. A lump on his head and some kind of burn on his wrist and ear. He looked ruggedly handsome but definitely not her style. What was it she felt? Her logical mind was at odds with her feelings. It had happened so fast. No future here, her mind told her. This is something special, her emotions shouted.

"I don't know. Well OK, but...I'm going away in the morning...I'll tell my brother I'm leaving."

They walked out together. She got in his car and they drove a short way to a small house.

"You live here alone?"

"Just me and my sister. I'm going away. She'll take care of the place while I'm gone. See you around." She opened the door and headed for her front porch. Jason got out of his side of the car and caught her by the arm before she reached the door. He turned her around and kissed her. She didn't resist. It was a long, slow kiss, gentle and unhurried. His hand slid up under her blouse and the bare smooth skin of her back radiated with the same heat he had felt earlier. He was overwhelmed with the new feeling. No bra, no resistance, no awareness of time as the seconds passed. He wanted the moment to last forever.

"May I come in?"

"No!" She turned and disappeared into the house.

He stood there in the darkness. Was it real or was it an illusion?

He got in his car. There was a partial six-pack on the floor in the back seat. His mouth was dry. He pulled a can of beer from the loop of clear plastic. It felt warm. He popped the top and took a long drink. It tasted different. The taste was familiar but different.

He started the car and pulled onto the road. As he drove past the old park where he and his son took their walks, an old memory came crashing in on him. It hit him hard. He pulled into the dark empty parking lot of the park and turned off the car. "Can't drive when you're time traveling," he thought.

He let his mind take him back. His sense of taste was doing the driving.

It was winter. He was fifteen years old and he and his friends were at the Little League baseball diamond. Leaves swirled in the eddies of wind that floated through the partially enclosed dugout.

His friend Gary reached into a carefully stacked pile of leaves and dragged out a case of beer. The weather was cool, but the beer was warm. Gary pulled out a church key and opened three beers, one for each of them. Jason had never drunk a beer before. They all sat around in a conspirators' circle and lifted the bottles to their lips. The semi-cool beer flooded his senses. It was a new experience.

A first. Never to be experienced again, for each new experience can only be experienced once. Or so he thought. Now he was unsure.

The darkness of the car raced in on him. The drink he had just taken. It had somehow been the same. The same as the first.

His mind raced back. It was all there. Had a single kiss somehow renewed him? Was he starting over? The friendship of childhood so pure, so real. Was this new woman comparable to his old childhood friends?

The experience in the dugout came to him again. He didn't want to let it go. He continued on his journey back to revisit an old chapter of his life. A most pleasurable chapter of childhood friendship and fun. The night full of 'present' quickly gave way to his childhood where love and friendship reigned. He didn't need a booth in some bar to help him travel back in time, he could do it right here in the car. His head rested back against the seat and Real Time faded into nothingness. The night quickly passed.

The sunshine startled him to consciousness. His back ached from the long night in the car. He started the car and headed home. He slipped quietly into his house and kicked his clothes off. He threw a quilt over himself and settled down on the couch. His son was up before he could close his eyes.

"Let's go to the park today." Jason reached for his clothes and before long he was back again at the park.

"Look at that stump, dad."

Jason walked over and looked at the old oak that had been hit by lightning. It had been removed. Nothing left but a stump and a few piles of sawdust.

"You could tell how old it was by counting the rings." Jason began to count. Over a hundred years old. The thought took him back to his thoughts about the past. Plowing the fields with a team of horses. What was it like then? He noticed a small oak sapling growing next to the old stump. What will life be like when that one is grown?

CHAPTER 10

THE MONDAY MORNING alarm was especially annoying on this new work week. The weekend went so fast and the week always seemed to go so slow. "If I could control time, I could make It go as fast or slow as I wanted to," Jason thought, as he headed for the old copper factory. How many more lessons do I need before graduating as a bona fide time traveler? The thought made him laugh out loud. Two lessons down, two to go. His life did seem to be changing though. It was becoming more difficult to tell the difference between fantasy and reality. Weekends like the one he'd just finished weren't helping his marriage any, either. Not that there was much left to help.

He had been telling himself he was staying in the marriage to hold on to his son. Now he wasn't sure. The fact might be he was just too lazy to do anything about the hole he'd dug for himself.

He pushed the uncomfortable thought quickly from his mind. He waved at the big security guard as he drove through the gate past the brick guard house.

He was early enough that he didn't need to stop his car at the door, punch his timecard and run back out and park. That was also new for Jason. He was beginning to get to work on time. He parked the old Pontiac and made a leisurely walk down the familiar factory aisle to the locker room. He quickly changed from his blue jeans to his work clothes. He grabbed the 'first stand' tongs that now occupied his locker. Ken Morehead intercepted him as he came out of the locker room.

"You'll be on the line today. Help the foremen give breaks to the stickers and try to stay out of trouble. Give me your first stand tongs and I'll take them to the Hill. Report to Fred and he'll give you a rundown on how to operate the line. We'll be rolling in about fifteen minutes. That should give someone as smart as you plenty of time of time to know everything there is to know about the equipment."

Jason wasn't bothered by the sarcasm. He was becoming used to the territory. Ken wasn't really singling him out, he was that way with everyone. He had been chewed out, but he hadn't gotten a written warning yet. If he were being set up to be fired, he would have been sent up to personnel for a written warning. It was becoming obvious that if you could do the work, you were needed on the Hill.

Fred was waiting for Jason at the end of the long water trough directly opposite the cafeteria door.

"How's it going Patch?"

"Not too bad. Guess I'm on the line today. Morehead says I have fifteen minutes to learn the equipment."

"There are four of us down here. If you get into trouble, we'll bail you out. This is the easiest part of the whole mill." The other two linemen were making their way to the line. Everyone seemed friendly and helpful.

"First, let's draw straws to see who buys coffee." Fred went over to an old broom, turned it upside down, and plucked four straws from the broom.

The other two linemen gathered around as if this was a familiar ritual. Fred broke one of the four straws and discarded half. He turned his back to the other men and spent a little time arranging the straws to be drawn. He turned back to face Jason and the other two men.

"You draw first, Patch."

He held the straws out to Jason. Jason looked at the broom straws between Fred's thumb and fingers. An innocent enough game. He could see four straws sticking above Fred's fingers and three straws sticking out below. Jason

scrutinized them closely, trying to determine which of the

four straws the short one was.

Was the short straw on the extreme outside position? Or was it one of the middle two?

"Come on Patch, don't take all day. We got work to do."

Jason reached out and pulled one of the middle straws. It was the short one.

"I drink mine with cream. Shorty and Junior drink theirs black."

Fred tossed the remaining straws in the waste basket and reached for a coffee carrying tray. He handed it to Jason. It had a handle made of solid copper bar stock, with number nine wire forming four loops, one for each cup. Jason headed for the cafeteria. As he walked away, he caught a glance that was being passed around between the group. A conspirators' glance.

As he walked to the coffee machine his mind raced over what had just happened. Fucked again he thought, but how? He'd seen all the straws with his own eyes. It was a fair game. He pushed the thought from his mind. Just Monday morning paranoia he thought to himself as he inserted the money into the coffee machine. He returned to the line with the coffee.

Fred was waiting for him when he returned. The others were busily getting ready for the day's run. He handed out the coffee. Fred took a sip of his coffee and began to explain the line.

"The finish sticker sticks to the finish mill and the copper rod travels down those long tubes to the baskets."

Jason followed Fred's line of sight to the four long tubes terminating at a huge piece of equipment that rested on tracks positioned above the four-foot-wide trough of quenching water.

"The copper rod is wound into a coil by the baskets. Then it drops to the cooling water. The conveyor finally brings the coil of black copper to us."

There were two hydraulic cutters hanging at the beginning of the line where two men were stationed to receive the copper coils as they emerged from the cooling bath.

Fred continued, "As the copper emerges from the water and travels to the roller conveyor, the man at the inside position cuts off about a foot of the bottom end of the coil. The man on this side cuts a foot off the top end of the coil. The ends are distorted and out-of-round from being slapped and whipped during the rolling process and the distorted part needs to be trimmed."

There were two scrap tubs next to each station for the deposit of the copper ends. Jason and Fred walked to the next station.

"These are the automatic tie machines. When the coil breaks this beam of light from the electric eye, a metal stop plate comes up and stops the coil." Fred passed his hand in front of the electric eye and a rectangular piece of metal rose out of the conveyor. Jason tried it. He broke the beam with his hand and watched the stop plate raise and lower.

"When the coil hits the plate, another sensor moves the tie arms into position." Fred placed his hand over the second sensor and the tie arms cycled in.

Copper wire automatically fed through the tie arms, formed a loop around an imaginary coil of rod and tightened into a smaller loop. The twister part of the mechanism twisted a knot in the number nine wire and sheared each wire thereby releasing it from the tie arms. The arms then returned to their original position.

"The coils continue to the end of the conveyor and are dumped off the conveyor onto one of three moveable pans. You dump fifteen coils to a pan, and then pull this lever to index the entire pan assembly."

Jason pulled down on the lever and the large metal pan assembly cycled one position.

"When two of the pans are filled, the overhead crane moves in and picks up the two rows of copper coils and takes them to the pickling operation."

Jason looked up at the large overhead crane. It was identical to the rod mill crane. Two tracks supported the large monstrosity as it traveled back and forth from the end of the line to the pickling operation.

He looked to the pickling operation. It consisted of a long row of six pits. The first pit contained acid.

The copper rod, blackened from the heat of the furnace, was allowed to sit in the acid for a while. The black scale, dissolved by the acid, was collected at the bottom of the acid tank via electronic plates. After the acid tank, the copper coils were moved to a caustic tank to neutralize the acid, and then to a rinse tank to cleanse it of the caustic solution. The two groups of three tanks had just enough capacity to keep up with the production of the rolling mill. The finished coils of rod were moved from the final rinse tanks to a storage area.

Fred moved back to the automatic tie machines and continued his explanation, "Occasionally these machines will malfunction, and you will need to flip the controls from automatic to manual."

Jason looked at the control panel. There was an off-on switch marked 'auto' and 'manual', followed by five buttons.

Fred switched the control panel to manual and demonstrated how each button controlled a separate function of the auto tie system. These controls were less complicated than the furnace controls. Fred quickly went through the demonstration and Jason watched patiently. Here, as up on the Hill, each second was important. The black copper coils came every eight seconds. The line was a part of the continuous operation. It reminded Jason of an old Charlie Chapman movie.

Charlie was working at the boxing operation in a hat factory. He would always get behind, but the hats would keep coming. These tiers were the same way. You weren't allowed to shut the line down if you got behind. If the tiers didn't function properly you could try to straighten them out, via manual operation. If you couldn't accomplish it by switching to manual and switching back to auto, you were forced to tie each coil by hand. If you couldn't keep up, the man at the end would dump the loose coils to the pans, and the coils would be re-tied by hand in a different area, usually by the guy that screwed them up. That meant when everyone else was on break, you'd be over in the re-work area tying coils.

Jason heard the long slow, rod mill whistle calling for the first bar of the new day.

Fred pointed to a dented rectangular box below the tie arm control panel. "I almost forgot to tell you. If the line shuts down, kick that fuse box below the panel. There's a loose fuse in there and sometimes it hangs up. Kick it real hard and the line will start up again. You start on the tiers."

Jason looked at the battered panel. It was obvious that it had been kicked many times. He wanted to ask Fred why they didn't fix the loose fuse, but Fred was already walking to the first station.

Jason heard Big Bertha groan as the first bar passed through her. He couldn't see the big mill, but he could feel her. He couldn't see the black man standing at the first station with his eyes, but he could see him with his mind. He could see them all, standing with folded arms, awaiting

the cherry red, copper snake as it moved ever closer to the Hill.

John made the first catch of the day and passed it to T-Bone, who in turn passed it to Billy Joe. Billy Joe passed the bar to the baskets via the long tubing that extended from the final mill.

Jason could hear the bar rattle through the tubing and into the huge coiling machine. He looked down the long trough of water and watched as the first coil released to the waiting quench water. A plume of steam rose from the water as the coil dropped into the trough.

The underwater flat conveyor slowly transported it towards the first cut-off station. The coil was submerged but its position beneath the water could be followed by the rippling wake it left on its journey to the front of the line.

As it emerged from the cooling water the two men at the first position cut the top and bottom ends of the coil and tossed the waste copper into the scrap barrel.

Jason waited patiently as the coil drew closer to his station. He was on his feet and prepared for action. The coil broke the electric eye beam, the stop plate emerged, and the tie arms cycled in. The number nine wire fed into the arms, tightened around the coil, twisted, sheared and returned to their original position. The stop plate disappeared into the conveyor and the coil continued its way to the end of the line. Jason watched as Shorty dumped the coil to the waiting pan. By this time the second coil was emerging from the water.

Jason heard the two short blasts from the rod mill whistle and listened as the rolling mill rumbled up to full speed. The whistle always made him tense. It prepared him for action.

He watched the second coil pass into the domain of the tie arms. It too was tied and continued its way to the end of the line. The coils were coming faster now but everything was running smoothly.

All keyed up and nothing to do. He walked down towards Fred and watched as Fred pulled the bottom end from beneath the emerging coil and cut it off.

"What am I supposed to do?"

"You're doing it. You just watch the tiers for twenty minutes and then everyone trades positions."

"You mean all I do is sit there by the tiers?"

"That's right. At the end of the first cycle you go to the Hill and help the foreman give the breaks. The assistant roller will replace you down here. When you're done giving breaks you'll come back to the line."

Jason couldn't believe it could be so easy. He liked this sitting around business. After the trials and tribulations on the Hill, this was a piece of cake. He went to the tiers and between bars he practiced turning the equipment to manual, feeding the wire through and then twisting it off. He quickly became very handy on the equipment.

It was simple compared to what he'd learned in the wire mill, the furnace and the Hill. After ten minutes he was becoming bored. He grabbed an empty waste basket, turned it upside down and used it for a chair. He leaned back against a vertical steel I-beam and propped his feet up."

What a deal, he thought to himself as he watched each succeeding coil pass through the automatic tiers.

He was about to drift off into a daydream when a sudden quietness startled him. The line had stopped.

He jumped to his feet and turned the tiers to manual. He frantically pushed each successive button. Nothing happened. Not only did the tiers not work, nothing was on. No blinking control lights the conveyor just sat there. Nothing was moving.

He was beginning to panic. He was afraid the stopped line would shut the mill down and he would somehow get the blame.

"Kick the box Patch!" Jason looked to Fred. He was pointing at the box below the tier control. Jason looked to the box. He looked back to Fred. It didn't make sense. The fuse box for the tier control shouldn't shut the whole fucking line down.

"Kick the som-bitch!"

Jason raised his foot and kicked. Nothing happened. It seemed the line had been down for an eternity. Jason expected to hear the shutdown whistle any second.

"Kick it harder."

Jason stepped back and gave a mighty kick. The entire tie arm assembly shuddered and miraculously the line came to life. The green 'auto-on' light blinked harmlessly, and the next coil rolled into the tiers and passed through to Shorty at the end of the line.

A wave of relief flooded over Jason. It didn't make any sense, but it worked. The line was back normal. Only seven seconds of real time had elapsed. It was time to go to the Hill. The Hill seemed a relief. It was a lot more work, but at least there were no unknowns. He relieved John at the first station. John stepped out and Jason stepped in. The next bar hit with a thud and Jason made the catch.

The day continued with little else happening. Just the continuation of repetitive work cycles. He realized that he'd learned the rolling mill. The great unknown had been learned and he was now a mill man. The seven years in the wire mill seemed to be part of a past life. There was no graduating ceremony. No certificate of learning just a feeling. A feeling of knowing ...knowing there was nothing new to learn, just a refinement of his skills...and then what? Jason didn't know.

By Thursday Jason had the line well in hand. This day had started like the others. Fred had held the straws; Jason had been the first to draw and for the fourth day in a row he had drawn the short straw. Fred again threw the straws in the waste basket and Jason went after the coffee. This was more than a bad run of luck.

This was some kind of conspiracy. It made him mad that he could not figure out the simple game. He had also noticed that the line shut down only when he was on the tiers. Oh sure, no one had to tell him to kick the box anymore. When the line shut down, he quickly jumped to the rescue and kicked the box, but it only happened when he was at that station.

Jason passed out the coffee. Everyone slapped him on the back and thanked him. He busied himself with the tasks of preparing the line to receive the first coils of the day.

Jason walked over to the line whistle and gave three short blasts, the signal for a rod mill electrician. Fred quickly came over to Jason.

"What's up, partner?"

"I thought I'd get an electrician and see if he can fix the loose fuse."

"I'm running the line down here. If anyone calls for an electrician, it should be me."

"That's okay with me, when he gets out here, you explain what's happening. I don't care who gets it fixed, I just don't want to force the mill to go down because of a stupid loose fuse."

Jason noticed that Ken Morehead walked out of his office and was standing on the high catwalk watching over the line.

Fred saw the superintendent; he also saw an electrician slowly making his way to the line. "No problem Patch. You called him; you talk to him."

Fred walked away.

Jason had bid on the electrician's job previous to the rolling mill, but he didn't have enough seniority to get it. He didn't know anything about electricity, but he knew most of the people who had the job didn't know much about it either.

The electrician's name was Clem. He was a heavyset fortyish man with graying black hair. His tools hung on a belt like guns in a holster. Like the rest of the electricians, his clothes were clean and crisp. If these guys ever did any work, it wasn't obvious by the way they dressed. 'A place for everything and everything in its place,' was the thought that came to mind as he approached. He walked up to Jason and rested his hands on his tools.

"You blow the whistle?"

"Yeah, there's a loose fuse in that box over there under the tie arms. Think you could get it fixed before the line starts up today?"

Clem looked to Fred and then to Jason and then up towards the Superintendent's office. Ken was still standing at the rail glaring down at them. Clem looked down at his feet as if searching for the right words.

"Ain't got time to fix it now. Mill starts up in a few minutes."

"How about leaving a note for second shift to fix it?"

Clem kicked at a piece of crud that was clinging to the sole of his work shoes.

"Actually, it can't be fixed. Uh, the tiers are obsolete, and they don't make that fuse box anymore. Fred put in for a new panel, but it won't be in for about a month. You'll just have to baby it along until the new panel gets here."

Jason thanked him and Clem walked away. Jason walked down to where Fred was standing. Fred seemed anxious to find out what had transpired.

"Well, is old Sparky going to fix it?"

"No, he said you already ordered a new fuse box last week. He said he and you had already taken care of it and as soon as the new box came in, he'd fix it."

"That's right. Of course, I usually don't consult rookies before taking care of business. If you have any more ideas, talk to me first."

The long slow blast of the mill whistle broke into their conversation and Jason walked back to his station at the tie arms. He glanced up and noticed Ken walking back to his overhead office. The tie arms worked fine for the remainder of the day. In fact, nothing of interest happened. When the final whistle blew to signal the end of the day, Jason loafed back to the shower room. He took his time in the shower and by the time he was dressed, only he and old Al were left in the locker room. Al was always the last one out. He looked the same at the end of the day as he had at the beginning.

He moved with the slow deliberate action of many old people, a carefulness that kept him a couple of steps behind the rest of the world. He looked at Jason with an air of suspicion.

"You're taking your time getting out of here. What are you up to?"

"Just checking on some riddles."

Al shrugged his shoulders and ambled out of the locker room door. Jason followed him out, but at the aisle that turned for the line, he slipped away. He walked to the tiers and fished in his pocket for a dime, found one and removed the two screws which held the fuse panel door closed. It had been dented from being kicked many times. The door was out of square and the hinges were battered and worn. He got all his fingers behind the lip and pulled it opened. It was empty. Just an empty box. He pushed it shut, replaced the screws and walked to the other end of the line. At the front of the line within reach of the man at the outside cutting station but out of view of the tier station, was the 'On-Off' switch for the entire line. Below the switch was another fuse box. Jason used his dime to unscrew the two screws holding the door closed. This fuse box contained three large 'tube type' fuses with brass ends.

Each fuse was held in place by two brass clips, top and bottom. He used his pencil to pry the top of one fuse from the top holder. He pressed it back against, but not into, the clips so it made contact.

With luck and a good jolt, like someone throwing the switch to the off position, the fuse might shake loose and break contact.

Jason closed the door and screwed it back in place. He walked to the waste basket, picked it up and carried it to a large garbage barrel. He carefully sorted through the day's trash until he came to the broom straws. It didn't make sense, there were no long straws, just seven short ones. Had Fred broken them all in half before the draw? He thought back. He watched his memory as Fred turned his back and arranged the straws.

"Fuck." Four days in a row and he hadn't caught on. He placed four of the short straws sticking out from his thumb and first finger and then positioned the remaining three short straws to stick out below his fingers. He was amazed at how honest it looked. He could feel the gap between the two groups of straws, but his eyes couldn't possibly see that there was no long straw to be drawn. He shook his head.

Jason replaced everything as it was and walked toward the long aisle leading to the time clock. Halfway out of the building he came across Al who should have been out by this time.

"What you are waiting for, Al?" Jason wondered if he had hung back and watched. "The men ever play any practical jokes on you, Al?" Jason knew it was a dumb question, everyone got their fair share of tricks played on them, some more than others.

"Sure, but I don't think much of them. Sometimes simple jokes escalate into wars. I don't think much of war, especially stupid ones over stupid issues. You'd be smart to Just ignore other people's stupidity."

Jason was sure he'd been caught. Al must have hung back and watched. What he'd done could be considered sabotage and that was a firing offense. "You ever get involved in horse play?"

"Not anymore."

Jason punched his card, and the two men went their separate ways. Well, he reasoned, if he'd been caught that was that. He could try to sneak back and fix the fuse, but if he got caught in the fuse box, he'd never be able to convince anyone he was trying to fix something. Somehow, he knew Al had watched him, yet he was also sure he wouldn't say anything. He'd just have to risk it.

Jason was in early Friday. He'd changed his mind overnight and was hoping to get in and fix the fuse before anyone else got there. Ken was already in the cafeteria when Jason walked up to the line. There would be no chance now. He walked in for coffee. Ken put money in the coffee machine for him. Jason pushed the button and watched the black liquid drain into the cup.

"Good to see you here so early. What happened, your old lady throw you out?"

"No, not yet I'm just trying to get employee of the year."

Both of them laughed at the remark. There was no such thing as positive reinforcement at this place. You were expected to do well. If you made mistakes, you got your ass chewed or worse.

"Where am I today?"

"On the line. We're ahead of schedule so far this week." Ken looked for wood and did a quick knock. "If everything continues to run smoothly, we'll be close to a record week. Al was in here earlier. He said he'd been watching you."

Jason's stomach did a flip flop. His mind started running through all plausible excuses he could use for being inside the fuse box late yesterday. Ken continued.

"Al says you may get better than T-Bone if you keep at it."

Jason let out a long slow breath. "It's coming. If I get the time and practice, I might be able to teach old Al a thing or two."

Ken just laughed. "That would be rich. If you do something Al hasn't seen, take a picture. I'd like to see it too." Ken walked out and headed for the office.

Jason walked out to the line and began getting ready for the day's run. He noticed the other linemen coming out of the locker room. The coffee carrying widget was sitting on the inspection table. Jason grabbed it and went to the cafeteria. Might as well make it five in a row he thought to himself as he headed to the coffee machine.

Jason brought the coffee to the line and passed them out.

"What's the occasion, Patch?"

"It's Friday. You guys have been so great to me all week I just thought I'd buy for the line." He slapped Fred on the back and headed for the tiers.

Ken was in his office talking on the intercom to both pulpit operators. He was aware of every day's production and kept a running total of tonnage in his head. There were no bonuses for record production weeks, but it was always a personal incentive for him to break the previous record. Something to keep him interested. To prove to himself that he was the best.

Ken pushed down the intercom button and talked into the mike. "We're on a roll this week guys. Sky, you check with the rear crane operator and make sure they have enough bars to get us through the day. Have Jake double check all furnace operations and make sure we don't have any stupid errors.

Fred, make sure the front-end tiers are perfect. Tell Shorty to have an extra coil of tie wire brought up. Find Al and put him on the line for extra coverage. I don't want any shutdowns due to unpreparedness."

Sky was also aware that things had been going perfect. The mill was a multifarious monstrosity. Seldom could you get through a week without something causing a shutdown. He pushed the 'talk button' on the intercom. "We got you covered, Ken. This day's gonna be perfect. I got the back end under control."

Sky released the office intercom button and pressed the furnace button. He watched as Jake moved to the intercom.

"Ken just told me we're rolling for a record. He wants Al up on the line for added protection. See if you can find him and point him that way."

Jake walked away from the intercom to the side door and out to the old rail spur along the side of the building.

There sat Al on an old empty reel. His back was against the building, his eyes closed. The sun was barely above the horizon and the tranquil scene was so pleasant Jake was hesitant to speak. He stood there for an exceptionally long minute.

Al broke the silence without even opening his eyes. "What's up, Jake?"

"Ken wants you up front on the line. He's looking for a record."

Al said, "He won't get it."

Al was a man of few words. Most of the time he communicated with a shrug or a nod. When he did say something it often didn't make any sense until later.

The simple prediction didn't make sense to Jake, but he believed it. It worried him that somewhere, something was going to go wrong, and Al already knew it. He was immediately concerned. Had he forgotten to prepare properly at the furnace? Had he missed some little item? If so, why hadn't Al told him?

"The problem ain't back here, is it?" asked Jake.

"No, everything's fine back here. It won't do us any good for me to go up front." He got up and ambled off towards the front of the rolling mill.

Jake watched him go and then walked back to the furnace. He didn't know how the old man seemed to know things before they happened, but it had happened too many times not to take him seriously. Not to worry, Jake thought. At least it won't be me that causes the shutdown. Jason saw Al walking toward the line and his stomach did another flip flop. He wanted to fix the fuse, but it was too late. If Al knew, why hadn't he said something? Was Al waiting for the line to go down and then nail him? Had Al come in early and fixed it himself? This worry was worse than working. Al walked over to the tiers.

"Morning Patch. Ken sent me over to help on the line. Guess we're going for a record today. How's everything going?"

Jason was sure he knew. He wanted to say everything was fine, but the lie caught in his throat. He searched for words, but none came to mind, so he just shrugged his shoulders. Wasn't that the way Al answered most questions? His response made the old man laugh. Al walked to the cafeteria and leaned up against the old space heater.

The Mill whistle gave out its long slow whine and everyone listened as the first bar passed through the first mill.

Real time was passing minute by minute just as it always did. Its' never-ending constant journey of semi-cyclic events rolled ever on to some unknown conclusion, but to Jason the minutes were beginning to seem like hours.

The first five minutes seemed an eternity. "Well," he thought, "let's do it." He turned the empty wastebasket over and sat down. Everything was running perfectly. He leaned back against the old I-beam, but before he shut his eyes he glanced into the cafeteria. There was Al, with a quiet smile on his face. It gave Jason a peaceful feeling.

Fred looked down to Jason and waited. "This rookie had been more fun than most," he thought to himself. Free coffee and free entertainment all week. Having a new man kick the empty fuse box was fun. The conditioned response was always a source of unbridled glee for the linemen. Sure, the kid would be sore as hell when he found out what was going on, but it was all part of the job.

It had happened to all of them. This rookie was cocky. This would teach him to be humbler around the senior mill men. He watched as Jason leaned back and shut his eyes. As he reached behind the panel to throw the switch, he noticed Al standing in the cafeteria. No matter, he thought Al never said anything. Al didn't like snitches any better than anyone else.

Fred slammed the switch to the 'off' position. The fuse separated a fraction of an inch away from the brass fingers. The line shuddered to a halt. Jason was listening intently. He had pretended to be relaxed when in fact, he was beginning to blush from anticipation.

He jumped to his feet and ran to the empty box below the tie arms. He gave a mighty kick. Fred was gleefully watching the action. He timed it perfectly. As Jason kicked, he threw the line switch to 'on'. Nothing happened. Jason stepped back and kicked again. Fred threw the switch again and again. Nothing.

Jason walked back and sat down. There was nothing to do. The line had a deathly stillness about it. Fred looked to the electricians' whistle, but Al was already there. Al blew three short blasts. Fred pushed the intercom button and told the pulpit operator the line was down. By the time he looked up from the intercom he saw Ken coming down the stairs two at a time.

Ken made it to the line before the electricians. "What the hell's going on?"

Fred had no time to concoct a story. "It wouldn't come on when I threw the switch." He blurted.

"Why was the switch turned off?" he yelled. Ken's face was turning a menacing deep purple.

The veins in his forehead were visibly throbbing.

"A tail got stuck in the rolls."

It was an obvious lie, but it was the best Fred could do on such short notice.

Ken looked to the rolls. There was no copper tail stuck in the rolls.

"Shorty got it out just before you got down here."

Ken looked to Shorty, who was nodding his head vigorously. He looked to Jason who was jacked back against the I-beam, as if nothing had happened. He walked down to the tie-arm station and Jason slowly got to his feet.

"What's going on, Patch?"

"I don't know, Ken. I kicked the box there, just like they taught me to do, but this time the line didn't come back on."

Jason was wearing his best dumb look. It had taken him many years to develop it and now it was much needed. It didn't work. Ken had known a lot of workers. From the almost retarded to the very cerebral. They all reverted to ignorance when there was a problem. No one ever knew anything. Not that he had been any different when he was a union man, but he was no longer a union man.

It was his responsibility as a Superintendent to make sure someone was punished for downtime. It didn't matter who fucked up. It only mattered that someone be hanged for the offense committed. The guy who ended up with the scrap or the one with the most unbelievable story was always the guilty party. He was fairly sure Jason somehow had a hand in this but as it stood, he would have to hang Fred.

Jason's mind was racing at exceptionally high speed. He could tell Ken was looking for a candidate for a hanging. He sat back down and waited. He was trying desperately to look calm and unattached. Suddenly his mind did a quick trace back to his old Geometry class.

The old school room was noisy as the first bell had not yet sounded. Jason walked up to the pencil sharpener and inserted his pencil. As he cranked the grinding wheels, he adjusted the pencil to achieve that perfect point, and looked out the partially opened window. It was a bright sunny day and it beckoned to him. Inside was all the boredom and work of another geometry class. The pencil sharpener was full, and Jason removed the cylindrical canister full of wood and graphite shavings.

The waste basket was below the table. Empty it into the basket is what he should have done, but the outdoors called to him. His moment's hesitation was all it took. He glanced around the busy, noisy room. No one was watching. Old man Garrett was nowhere to be seen. He looked to the room, back to the door, back to the window and with a flick of his wrist, the pencil sharpener canister flew out of the window and into the hedges. He glanced around again and made sure no one was watching.

Jason walked back to his seat and sat down. He opened his book to the day's lesson and tried to concentrate on the multitude of geometric shapes and numbers. He couldn't help but smile inwardly about the perfectness of the crime.

Old man Garrett came in the room and the class quickly quieted down. Jason was watching him intently to see if he noticed. To Jason, the naked pencil sharpener stood out like a sore thumb. He couldn't believe no one else had even noticed.

The lesson began. It was something about two points on a Cartesian plane. Garrett was drawing on the board, his back was to the class and all of a sudden, the chalk stopped. Frozen forever in Jason's mind. Jason somehow knew he was caught. The old man didn't turn around. He didn't look at the pencil sharpener. He didn't look out the window. He didn't even raise his voice, "Jason, would you go down and get the pencil sharpener canister." He continued his graph.

Jason was stunned. A few girls giggled but no one else seemed to know what was going on. He got up and headed for the door. He retrieved the canister, put it back in place and returned to his seat. Nothing more was ever said, yet it was never forgotten.

The present raced back in on Jason. Ken stood watching him. How long had he been gone? Probably only a few seconds. Did the Ken know? Ken turned and walked to the control panel where the electricians were opening the door of the fuse box.

One of the electricians pulled a pair of pliers out of his holster. He reached into the box and snapped the loose fuse back into position. He shut the door, threw the switch and the line came to life. "Just a lose fuse, Ken."

Ken looked around furiously. He wanted something to throw. He wanted someone to choke. Everyone took a step back. They watched as the fury came and went. He looked to the electrician first. "That's bullshit," he screamed. "Those are held in tight. They can't just get lose!"

The electricians turned and walked away. Ken went after them. "Don't turn and walk away from me," he yelled, "I need some answers!"

One of them turned around. "If you got something else for us to do, let's hear what it is. If all you want to do is yell, tell it to our boss." They turned and walked away.

Everyone else was busying themselves with getting the mill going. Each one was trying to stay out of the way. Ken hadn't found his target and was looking for a probable candidate. He fixed his stare on Fred. There was no place for Fred to hide.

"Patch claims he kicked the box like he was trained to do, and the line didn't come back on. This wouldn't have anything to do with playing tricks on green operators would it?"

It was a stupid question. Everyone knew the green operators were trained to kick the box. It was a kind of hazing. It had been done to everyone, including Ken. But Ken wasn't really asking a question, he was laying a trap and Fred was the prey.

Fred thought a minute. He had already made one blunder by admitting he had thrown the switch. It was bad luck the fuse had come lose at that particular time...or was it? Fred quickly glanced down at Jason. Jason met his stare. "Fuck you, Fred" was what the stare was saying.

Ken followed the instantaneous communication. It was all too obvious. This new kid was dangerous. Ken was also caught in the trap. He was forced to hang an innocent man. No matter though, no one was completely innocent.

Fred finally answered with a statement. "You want to talk to me; you go get my union steward!"

Ken glanced over at Al and noticed the slight smile on his face. He looked like a kid watching a cartoon on TV. Somehow old Al knew what was going on. He probably knew before it even happened. The thought enraged Ken even more.

"You fucking guys are like a bunch of kids," Ken yelled to no one in particular. He looked back to Fred. "On your way out tonight stop at personnel with your union steward and pick up your written warning for horseplay." He looked to Al. "Come up to my office."

A half hour of downtime. A thousand dollars' worth of scrap. No record run for the week. Was he the only one who cared? The mill was back up to speed before Al got to the overhead office. Al walked in and sat down. Ken lit a cigarette and leaned back on the old roller legged chair. "It was Patch who set Fred up and caused the shut-down, wasn't it?"

Al responded, "I remember a couple of times you shut down the mill because of horseplay. You remember the time you dropped the live cigarette butt in Shorty 's back pocket? Every time he leaned back against the mill it burnt his ass.

"Shorty kept turning around and feeling the mill cap, thinking that was where the heat was coming from. By the time he figured it out, he had to drop his pants. He missed two bars in a row, and we had to shut down."

Ken thought back. The vision of old Shorty dropping his pants and slapping the heat out of his flowered boxer shorts got him to laughing.

"Those were some fun times," he said. "Things are different now though. We run a lot harder and the pressure is on me to get the production out. Hell, in the old days we used to..." He stopped in mid-sentence, remembering the chore he had to do. The old times didn't matter. He wasn't a union man anymore. He was responsible for taking care of the problem at hand. He had to make all the participants pay, one way or another.

"You draw up an electrical sketch to move the line switch up to the pulpit area." He picked up the phone and called the boss in charge of the electricians. Al could hear Ken's side of the conversation.

"I want new electrical conduit run from the line to the front pulpit. Al will have a sketch for you by noon today. When I come in Monday, I don't want to see the old control switch in the hands of those idiots on the line. I don't care if you must work your whole fucking crew overtime. I want those controls changed by startup Monday morning."

Ken hung up the phone and called Komfer. "Get a written warning ready for Fred Mahorn. Put down horseplay for the reason. No details. If we give details, he'll just lie out of it... I don't care how much he fucking whines about it, I want it in his file for at least a week. It will take those lazy union fuckers at least that long to present a written grievance against the company. After they present their grievance you can make a trade, but you can throw it away only after they present their written grievance." He hung up the phone. "The real problem Al, is, what the fuck is he going to do next?"

"No one knows the future, Ken, but you know he can't do anything we haven't seen before." Al got up and moved slowly for the door. It would take him the rest of the morning to get the electrical sketch ready. He smiled as he shut the door behind him. It had been worth it! Al remembered the old days when he himself was green. At that time in his life everything new scared him. Each mistake was an adventure in learning. Now, it was so seldom that anything new happened, he genuinely enjoyed the break in monotony.

Chapter 11

Jason was home at his usual time Friday afternoon. His wife was waiting for him in the living room. What he wanted most was to grab a bite to eat and get out of there.

He looked around for Brian, but he was not in sight.

"Where's Brian?"

"He's at Mother's."

Jason headed for the bedroom to change shirts.

She stopped him with a statement. "Come here and sit down for a minute, I want to talk to you."

This was unusual. They never talked, other than what was necessary to get through the day. He walked over to the couch and sat down. I hope this doesn't take long, he thought.

"It's not working, Jason." She was not looking him in the eye, but that was not unusual, they seldom made eye contact anymore.

"Fuck!" he thought to himself. "What is it now? The washing machine, the garbage disposal, the car?"

"What's not working?"

"The relationship."

Jason was taken back. This was strange. They never talked about their relationship. He remembered when they used to talk, but now it always ended in a war. They didn't agree on anything. They never settled anything. Each issue had been discussed until they both became weary discussing it. Each war was fought to a draw with the final truce being they just didn't talk about that particular subject anymore. There was little left to discuss. They were so far apart in their thinking they had little in common. Now, all of a sudden, she was wanting to talk about the relationship, and the topic was so enormously hopeless that he was at a loss for words. Something was up. He leaned back and waited. The entire situation was impossible, but he'd come to accept it. He had decided to stick it out, to make the best of a bad situation. Something was about to change.

"I filed for divorce last week. You'll be getting a restraining order tomorrow. I thought you may want to get your things and go before you're evicted."

Jason physically slumped down. It was such a shock. He was flooded with conflicting emotions. Evicted! How dare she even think of evicting me! I worked seven fucking years in the factory to pay for this fucking place. His thoughts were racing, but he kept his mouth shut. Another voice from somewhere was saying. This is what you want. Haven't you wished her gone a hundred times? Haven't you anguished over your lack of freedom? His anger rushed in on him. Evicted? I can grab her by the throat right now and put her ass in the street! He waited for a long moment.

"What's a restraining order?"

"It's an order that restrains you from coming near the property. You aren't allowed to come in the house or call or in any way bother me or Brian. Do you want to leave now, or do you want the sheriff to show you the door tomorrow? You can use the car to take your things wherever you go, but the order says I get to keep the car. You'll have to bring it back."

She had a certain smugness about her, like 'last tag asshole', or 'Game Set Match, I win'.

Was that what this was, a game? Was marriage supposed to be a competition? One in which each partner strives to gain an advantage over the other through a constant barrage of arguing and fighting?

He thought back to his parents' marriage. He couldn't ever remember them fighting. He could only remember love and companionship. How far short he had fallen from that example. How miserable a failure his marriage had been? His mind raced back to the first time he'd seen her.

She was walking down the crowded aisle between classes in high school. He remembered the striking yellow blouse. He remembered the flash of emotion, the desire, the need. He had pursued, conquered, and…lost.

Lost is where he had arrived. What to do? What were his options? He walked to the phone and called his mother.

"Mom? Is it OK if I stay with you for a couple of weeks?... No, no problem, it's just not working...Thanks, good-bye."

Jason went to the bedroom and packed a small suitcase. He grabbed a handful of clothes from the closet. He looked at his possessions. Seven years in the factory, what had he acquired? A handful of hangers and a suitcase full of clothes. His old bicycle was in the garage. He would bring the car back and take the bicycle. The bike reminded him of Brian. The thought of not being able to see him or talk to him rushed in and threatened to crush him. He headed for the door. Somehow spatial realities seemed distorted. Time seemed to have stopped and, in its place, stood a vacuum, an emptiness, a void.

"After you unpack, bring the suitcase back with the car. It belongs to me."

His eyes flashed an anger she had seldom seen. Had she gone too far? She felt a sudden primeval fear. It started somewhere deep in her stomach and reached up to her throat. She was going to add a few final digs, but the words stuck with the fear somewhere in the deep recesses of her mind. She was silent.

The urge was there, and it was an awesome terrible feeling. She was in reach. He had the power. He felt it in his being. The power to reach out and kill. To choke until he felt the rattle. He fought it back, turned and went out the door. It was to be their last conversation.

An attorney contacted him through the mail and set up a hearing. Jason showed up at the courthouse and went to the proper chambers.

Her lawyer was tall and handsome. Wavy brown hair, clean fingernails, well-groomed. Slick was the word that came to mind. He looked at the lawyer and at her. No need for conclusions. It was obvious there was more here than a professional client-attorney relationship. The whole scene threatened to overwhelm Jason.

The attorney stuck out his hand. "Mr. Strider, I'm your wife's attorney."

Jason gave him a long look. He knew he was in over his head. He knew nothing of the law. A rookie again in a game of professionals. He was sure he was in for a loss. This Mr. Slick was about to bury him. He'd already denied him his son.

Jason refused to take his hand. In Jason's world a handshake was still used for friendship. It was impossible for him to show such a gesture.

Mr. Slick gave him a hard look. "Mr. Strider, you can make this as easy or as hard on yourself as you like. It doesn't matter to me. You'll pay one way or the other."

Jason just turned and walked away.

What followed was an example of Indiana justice. A lesson on how to twist the legal system. Mr. Slick waited for one of his friends to be appointed Judge Pro Temp. He called for a hasty hearing without notifying Jason. He got a judgement of maximum payment per week and then began the delays. Since Jason was unable to pay the full amount, the remainder was subtracted out of his home equity until it was gone. As there was no more to extract from Jason, Mr. Slick arranged for a final termination of the marriage.

Jason had not fared too well during the interim. He no longer stopped at the Old Mill tavern. His lessons in time travel had come to a halt. He visited most of the other bars in Fort Wayne, and as many other women as could be found. He had purchased an old GMC Van from his brother. The in-line, single-barrel, six-cylinder engine provided all the power he needed to transport him in his ever-thickening haze.

He seldom made it home to Mom's. The old van provided both transportation and a place to sleep. It was late summer before he reached his breaking point, not that one point was much different than another for a man lost in a time void.

Jason was driving down the road on his way to work. His head was pounding from the excesses of the night before. His clothes reeked of cigarette smoke and too many days and nights in his home on wheels. The depressing side effects of alcohol combined with the thoughts of his son allowed his anger and furry to surface. The past weeks of hatred and self-pity had isolated him from his old group of friends. The loneliness and pain welled up and he could no longer hold it in.

He screamed with all the fire and anger within.

Screamed at the inside of the old dingy windshield. He shook the steering wheel with unbridled rage. He screamed again in awesome anger. There was release there. He screamed again and listened as the sound reverberated off the tin walls. The pain brought both quietness and tears. He pulled off the road and waited. He knew he was crazy. Either crazy or awfully close to it.

A sane man doesn't act like this, he said to himself. Does life have to be this hard? Somewhere a voice answered him with a question, "Why give the world that kind of control over you?" From those words he found a thread of reason. I'm not going to give it control. I'll no longer respond. When he got to work, he was exhausted.

Big John was waiting for him in the cafeteria. He put money in the coffee machine and pushed the button.

"You look like shit, Jason. Don't you sleep anymore?"

Jason just shook his head. "I sleep plenty, I just don't seem to rest," He said in a rasping voice.

John had seen all the symptoms before. Newly divorced, coming in smelling like a stale bar, slow reflexes and paranoia. Either they pulled out of their tailspin or they didn't. Be a shame to lose this one though. He'd gotten better every day; despite the abuse he'd been putting his body through.

"Get through today and we'll be having two days off. You gonna drink up this weekend or you gonna find something positive to do?"

"To be honest with you John, I didn't even know it was Friday. I haven't made any plans."

"You'd be smart to lay of the booze for a while. Find yourself a friend that don't drink so much."

"Yeah, maybe I ought to travel back in time to when I was in the wire mill, things didn't seem so confusing then."

"That also fit the profile," John thought. "They start wanting to go back to the past, and they do their escaping in a bottle."

"Maybe yous ought to start thinking about the future and leave the past alone. Looks to me likes the past been beating you pretty bad."

Suddenly the thought of past and present reminded Jason of a four-fingered traveler named LeRoi Verita. How long had it been? How could he get a hold of him? It would be nice to talk to someone. To really talk. Not just a cheap hustle but a real conversation.

"Guess I'll just try to get through today. Who knows what will happen in the future?" But something deep within gave him an answer. LeRoi knows. If he could find him.

Chapter 12

LeRoi had been busy. He was always busy with some new project or theory. The first week Jason hadn't shown for his lesson had been disturbing. It was such an adventure to teach, for teaching was one of the best ways to learn. To try your ideas out on someone other than yourself.

He had done a lot of teaching at the local college, but it was the conventional type. The students paid their fees and LeRoi would feed them their daily ration of information. It was more like slopping the hogs than enlightening the souls, but it paid his meager way. Teaching Jason was more than teaching, it was an adventure.

The second week Jason hadn't shown and by the third week LeRoi was only doing a walk-through. He'd walk through the old bar and, not finding his student, he would retreat to the blackness of the river and do his own traveling.

He found he could travel with the poets and the authors of any time period. Grab a book from the college library, read the pages and disappear in whatever era was explained. He realized his time was running out and he should be doing something with his life but what it was he couldn't quite figure out. He had seen through a tiny window to his future while alone in the mountains, but it all seemed so long ago.

With Jason, he had thought he might be able to see more of the elusive future but now he was back to his patchwork routine of short-term temporary gratifying little escapes into other people's pasts. 'The Man from La Mancha' had been his most recent adventure, but each Friday night he hoped for more. He pulled his canoe into the soft sand at the edge of the old river and began his weekly ascent to the Old Mill Tavern. He walked up to the bar and ordered an Ouzo. Frank slid the shot across the counter and took the money. LeRoi walked around the partition and glanced to the worn Naugahyde upholstered booth. There sat his friend. Jason looked older. Older and more worn. Something like battle fatigue. Jason looked up from his drink and saw LeRoi. Just a glance. But the glance erased the void. It brought them back to where they had left off, like there had been no interruption.

Jason wondered why he had stayed away in the first place. He didn't know and it didn't seem to matter. They were here for the moment, and maybe this was the only moment that existed.

LeRoi walked over and sat down. "You buying or should I buy one for the prodigal son?"

Jason laughed. "You buy, my funds seem rather limited these days. What is a prodigal son?"

"Just an old story from the Bible. A man's son was gone for a while and when he returned the father put on a big party."

Jason thought about his son and the thought began to drag him down from his isolated moment of enjoyment and pleasant quiet surroundings. LeRoi watched Jason as the myriad of emotions flooded through him. Up one instant and plummeting back into despair the next. He was like a yoyo. There could be no teaching unless he could get past whatever was killing him. He didn't want to play the role of counselor, but he didn't have much choice, someone would have to explain the facts of life to him sooner or later. "What's eating you, Jason?"

"It's the cocksuckers' downtown." Jason said with conviction. "First they set up an illegal hearing and stole everything I ever worked for. Now they won't allow me visiting privileges for my own son because I'm behind in my support payments."

Jason sounded like a little kid whining because he'd gotten a bad call in a ball game. LeRoi might be able to help him but it wouldn't be easy. First, he'd have to educate him about the system.

"It's a voluntary system, Jason. If you don't like the rules, quit the game."

That wasn't quite what Jason expected. He had hoped that at least LeRoi would understand.

LeRoi continued. "What does anyone with courage do in a hostage situation? When the bad guys request money, they refuse to pay. Besides, he's not 'your' son."

Jason's head snapped up from its slumped position. His eyes shifted from self-pity to anger. He glared across the table at LeRoi. "You don't know me, you don't know my son, and you don't know my ex-wife." Jason stopped and thought for a second. He wanted to tell LeRoi he was full of shit, but he felt like he was walking into a trap.

He thought about the pictures of himself when he was five. They were identical to young Brian's pictures. Brian was his son. He knew it from his mannerisms, his heart and his soul. This time LeRoi was plain full of shit. He continued, "How can you sit there and talk of something you know nothing about? This time you don't have any idea what's going on." Jason was angry. He would defend his position to the death. He had never been more right about anything in

his life.

LeRoi took another sip of Ouzo. He looked across the table at the heat and fury radiating from Jason. Even if he were blind and deaf, he could have felt it. He waited and let the heat subside a bit. He needed to take him another step down the road of his ignorance before trapping him. "Who owns you Jason? Does your wife own you?"

"No!"

"Does your wife's lawyer own you?"

"No!"

The anger didn't bother LeRoi in the least. In fact, he almost seemed to enjoy it. He was so calm, like this was the most normal conversation in the world. A few of the patrons were beginning to look their way. They could feel the fury. Like before a fight starts, they were beginning to watch out of the corners of their eyes. Jason didn't care about the other people in the bar. He was in defense of his son.

"Does your mother own you?"

"No. No one owns me!"

"Does your father own you?"

"If he were alive, would he own you?"

This was stupid. How many times did he have to tell him? "Look. No one owns me! I'm my own man. I'm a sticker in a rolling mill! I make my own decisions. I go my own way. I buy my own food and I'll kick anyone's ass who fucks with me. Why are you doing this?"

"If your father was alive, would he own you?"

"Dammit! Nobody owns me! Nobody will ever own me!"

"Do you own Brian?"

Jason stopped. He felt the jaws of the trap slamming shut on him. He wanted to shout something, but the words stuck.

"No one owns you, Jason, and you can't own anyone. Not even your son. He belongs to himself. Even at his tender age, he is his own person, just as you are your own person. "Sure," LeRoi continued, "he is from you and he loves you as you love him, but one person cannot own another. You should have learned that simple truth long ago."

"Because you don't own him, no one can take him away from you. Just because they restrict you from seeing him in person, doesn't mean they can keep him from you.

"Tomorrow when you wake up, write him a note. Tell him you love him. Tell him he'll always be your son and you'll always be his father and when the time is right, you'll see him again.

"Tell him until then you'll continue to write him letters and when he has time, he should draw you a picture and send it to you."

Jason felt light-headed. He suddenly saw himself wrapped in chains. He watched himself stand and the chains fell away. Without them he almost floated off the floor. He looked across the table at

LeRoi. All Jason's anger was gone now. How could a few words set him free? LeRoi watched as the tension drained out of Jason.

"We put ourselves in chains, you know. We blame other people but we either do it to ourselves or we give other people the power to do it. It's always a voluntary system. To free ourselves all we need to do is arrive at the proper level of understanding. To achieve freedom, you only need to have the right thoughts. Haven't you ever heard the poem? 'Two men looked out from prison bars. One saw dirt, the other saw stars.' It's the same with all of us. Without the proper teaching and understanding we all build prisons for ourselves. No amount of money or constitutional amendments can make foolish men free. No bars can imprison an enlightened soul. Now, are you still interested in finishing your lessons in time travel?"

Jason looked around the bar. Everyone was back to normal. They were all drinking and laughing as if nothing had happened. To him the last few moments had been a roller coaster ride. Now suddenly, his roller coaster car had jumped the tracks and he was up in the clouds somewhere, floating, free of the world's influences.

This four-fingered fellow across the table had enlightened him so he could free himself and was now asking him to continue a lesson as if nothing had happened. Jason reached up and lightly slapped himself. He was sure he was in another daydream.

"There's no reason to check on yourself, Jason. You were gone for a while, now you're back, just like the prodigal son. Let's continue, do you remember the earlier lessons?"

Jason had to concentrate. It all seemed so long ago. "There are two types of time, Real Time and Mind Time. Usually they are in sync with one another but when you speed up Mind Time, Real Time appears to slow down. When you slow down Mind Time, like when you go to sleep, Real Time appears to speed up."

LeRoi said, "That's great, please go on."

Jason said," There are two ways to travel back in time. If you want to take your body with you, you can go to any part of the world where different cultures are still evolving. You can go to an Amish community and go back one hundred years, or you can go to the land of the mountain gorillas and go back a million years. If you want to travel back to a previous time within your own childhood, you go in your mind, your imagination. You can watch it unfold like a movie or a picture, but you can't change it. There is no way to change the past."

"That's exactly the way I taught it to you." LeRoi said. LeRoi wanted to tell him there was a way to change the past, but at this point it would confuse him. He let the lie stand. "A little lie told in the advance of information was OK, as long as you corrected it later," he thought. Of course, one was never sure if there would be a later, but that was a chance he needed to take.

Jason said, "The present continues but all of life's knowledge can be reset to zero with a kiss. Then you can begin again."

This time it was Jason's turn to sit across the table and watch the multitude of emotions run through a person's mind.

LeRoi straightened. That didn't come from him. His mind was doing a frantic search through the old classics, looking for a benchmark from which to speak but he found none. It made sense, but he'd not thought of it before. The only thing which came to mind was the old fairy tale about the frog and the princess. If you could reset everything to zero, you could re-experience many of the simple pleasures, maybe all pleasures. LeRoi said, "It was the dark-haired woman you met here wasn't it?"

"Yes. But I only saw her once. I've tried to contact her but she's out of the country. That one kiss started me on a very enjoyable childhood memory. I played a whole summer of baseball in one night. It was the most fun. I figure if a kiss can do that for me, think what a whole night in the sack might do. I might be able to transcend the entire universe."

LeRoi laughed. He was a bit envious. Since his return to civilization he had been lonely. There had been some women but only when he wanted to fulfill his physical needs. He had never met anyone to share his spiritual needs.

"You're probably onto something with that kiss business, but I'll need to do some research on it." Not that he would know where to go to look for information. Maybe he would just go to the library and look up 'Kiss for a New Universe' or In the 'Beginning There Was A Kiss.' LeRoi pulled out a small tablet and jotted a few quick notes and then put the notepad back in his pocket.

The lesson tonight was about CB radios and LeRoi began it with a question. "Do you know anything about CBs?"

"Ten-four good buddy." Jason said, I know a lot about CBs, he used them on vacations to avoid the police. He didn't see any immediate connection between CBs and time travel, but then he hadn't seen the connection with baseball, poetry or any of the other beginnings in previous lessons. One constant seemed always to be present. External influences began to disappear. The rest of the bar, even the rest of the world, seemed to slip away.

"Tell me what you know about CBs?"

"You turn on the two-way radio, adjust it to Channel 19, and when you want to talk, you push down the send button and talk into the mike. If anyone is out there, they talk back."

"That's correct. I mean, how does it work? How does the message get to the other person's ear, and how are you able to hear what they have to say?"

"It's like the telephone except there are no wires."

This lesson was starting out just as boring as all the rest.

LeRoi continued. "There are two parts to the two-way radio, a transmitter to send messages and a receiver, with an antenna, to receive messages. Why is your CB different than your car radio?"

"I can only receive on my car radio. I can't send any messages back to WOWO."

"How is your television different than your CB?"

"Same answer, I can receive but I can't send. They are both one-way pieces of equipment."

"Do you believe in CBs?"

Jason was getting that bored look again. "I don't understand the question, I just told you I have used them before."

"Do you believe they exist? Do you believe it is possible to have two-way communication without having wires attached?"

"Of course, I believe it. Everyone believes it. They even make little walkie-talkies for kids to play with. Everyone knows about two-way communication. All you need to do is tune your equipment to the same station or channel and you can talk all day long. What's the big deal about that?"

"It's no big deal. We both have two-way radios built into our minds. We also use those all the time. When we are both tuned to the same frequency, we can communicate without speaking, like with a glance."

"I don't believe it! You mean we could just sit here and talk to each other without saying anything? If that were true, why are you talking to me right now? Why don't you tune in to my channel and down-load whatever information you want me to have?"

The resistance was always there. From the beginning of teaching, each person was taught all the mechanical communication skills and taught to ignore or repress those skills which could not be explained.

"If we both had the proper mind set-up, compatible sending and receiving equipment, and if we both knew how to use it, we might be able to do just that. The way it is now, we are like children playing with two-way radios. At first neither knows how to use them. Without instructions, you fiddle with the controls and occasionally you may receive a message from someone else, but not knowing exactly what is happening you may misinterpret the message, or worse, never understand how to send out your own messages."

Jason was beginning to like this. He looked across the table at LeRoi. "What you're saying is that we all have two-way radios built into our heads, and you have the instructions."

"No! **You're getting ahead of me again. What I would like you to do is to accept there is such a thing as wireless two-way communication. That there is a possibility we have the ability to practice**

non-verbal transmissions. If you believe in CBs, you may be able to believe in this line of thought."

"Ok! So, I believe it's possible. What good would it be?"

"Well, for one thing we could go on time journeys together. I could go with you, or you could go with me. It could also be used to tell the future, at least part of it. It could also explain Deja vu, reincarnation, and premonitions. It would explain all those and more, but it's just a theory. Because a theory fits well doesn't mean it's true, but it's a lot easier for me to believe in CBs than it is to believe in all this psychic phenomenon you hear about all the time."

Jason was becoming more and more confused. "I'm not following you. First we're talking about CBs and then all of a sudden were talking about a whole group of things that no one really knows anything about."

"That's correct, but if anyone had it figured out, it would be common knowledge, wouldn't it? I just wanted to give you the basic thought and let you think about it for a while. Oh yes, one other thing. Did you ever notice the big CB antennas on the really good CBs?"

"Yes."

"Do you know why they're so long?"

"No."

The best units have antennas the same length as the wavelength they are receiving. The transmitter sends at a wavelength of about one meter. That's the same

length the antennas are, one meter."

Jason didn't understand but he was willing to go along with this part. He hadn't understood some of the other parts at the beginning, but the pieces seemed to fit together.

LeRoi continued, "Have you ever had thoughts pop into your head and had no idea where they came from?"

"Sure, it happens all the time."

"Well, where do you think the thoughts come from?"

Jason had to think about it. "Well, if what you're saying is true,

I could be picking up information from anyone, at any time. That wouldn't work though because there are so many thoughts going on all the time. You would be barraged with everyone's thoughts. It would drive you crazy in about ten minutes."

"Not if your receiver was very selective. You would only pick up the people who were exactly tuned to you, either by birth, love, or coincidence. It could also be very selective with your mind filtering out all extraneous messages."

"Can you do that?", asked Jason.

"Well, no, not on command, said LeRoi" " But I know there are such things as cause and effect. If something occurs there is a reason for it. Until any phenomena is understood, there will be many theories about the effect. Until someone proves something, you're stuck with accepting whichever is the most logical explanation. I don't want to get involved in any more of the specifics right now. Right now, I'd like to talk about traveling to the future."

"OK Mr. LeRoi Varitia, tell my future for me. But only tell me a couple of days in advance. That's safe enough isn't it?"

"What time do you set your alarm in the morning?", asked LeRoi.

Jason replied, "I have to be at work by 6:48 am, so my alarm goes off at 6:00 am."

"OK", said LeRoi. My first prediction is that at 6:02 AM on Monday morning you will be standing in front of the ivory throne with your dick in your hand."

Jason laughed. "What kind of a prediction is that? Anyone could predict that event. Besides, I could prove you wrong by changing my alarm by fifteen minutes. Then your prediction would be wrong."

"That's exactly the point I want to make. If you can see the future, you have the ability to change it. Therein lies one of the most valuable lessons in time travel."

"Why is it valuable? "asked Jason.

"If you're a baseball coach and you're down two runs in the first inning, you look eight innings ahead and realize that if the trend continues as it is, you're going to lose the ballgame by sixteen runs."

"The good coaches make strategic moves to change the predicted event. If they choose correctly, the foreseen future is changed and the outcome swings to their advantage."

"Ok, you're saying baseball coaches are time travelers. Do they know they're travelers?"

"Some of them know it."

This was becoming more interesting. If he were a traveler, maybe he could change professions and get out of the factory. "I see what you mean, but it doesn't always work, does it?"

"No, nothing is foolproof, but you can see the possibilities?"

"What other ways are there to see the future?"

"There have been many prophets and seers who could see a window into the future in their dreams. Haven't you ever heard the saying, 'Death is but a door, Time is but a window.' The ones who understood what they were seeing became famous or were burned at the stake.", said LeRoi.

"For example?", asked Jason.

"In the Old Testament there was Daniel. He not only could see the future; he could also see the future through other people's dreams."

"You mean those people were time travelers' way back then?"

"Yes, but again, let me caution you. The ones who really understood, quickly found out how dangerous it could be. Daniel also had a run-in with some lions. Being able to see the future is one thing, knowing what to do with the information is quite another. It's kind of like discovering the answer to a riddle, but not being able to tell anyone."

"I don't understand. What's the big deal? I mean if you see the truth, then why not reveal it? What can the truth hurt? Wouldn't everyone want to know it?"

"It has to do with the Heisenberg Uncertainty Principle."

"What is the Heisenberg Uncertainty Principle?" asked Jason.

"Back in 1923, a German named Werner Heisenberg was working

on a theory developed by de Broglie. Heisenberg, Schrodinger and many others were attempting to put together a completely new theory called wave mechanics. They had been trying to look at very tiny particles so they could predict what these particles would do under various groups of

experiments.

"Heisenberg realized that in order to observe what these particles were doing; he would need to shine some kind of light on them so he could see them. The problem was that these particles were so small that the force of the light hitting them would change their natural state and you could not predict where they were going because of the light. Henceforth, the Heisenberg Uncertainty Principle."

Jason was beginning to see the connection. "What you're saying is that if you see the future, you can't say anything about it because it will change it."

"Bingo! That's it in a nutshell."

Jason saw an immediate problem. "That's not any good at all. What you're saying is if I get good enough to see the future, I can't tell anyone because me telling it, or shining a light on it, will change it and I'll look like a fool."

"Yes, that's a problem, but it doesn't mean it's not a valuable skill to have. What it means is, if it's misused, you'll make a fool of yourself. Not much different than any other gift, if they're misused, they become negatives instead of positives."

"Was that Heisenberg fellow a traveler?"

"I don't know, if he was, he didn't write anything about. At least not that I know of. But because of the problem you stated there have probably been many travelers who kept their thoughts to themselves."

Jason was thinking of all the times he had heard of people telling the future. "What about the lady that foresaw her plane crashing and didn't go on the flight?"

"For her, that was a very practical application of what we're talking about."

"But if she knew it was going to happen, didn't she have an obligation to call and tell the airline it was going to crash. She could have saved all those people's lives."

"If it's the same story I'm thinking about, she did call and warn them. The problem was the airline can't cancel every flight due to crank phone calls. Besides, if they cancel the flight, the plane doesn't crash, and the lady is wrong about her prediction."

"That's simply great. If she's right and anyone listens to her, her prediction is proven to be false, but if she's right and doesn't tell anyone, a bunch of people die."

"She didn't die."

"What should she have done?"

"She should do whatever everyone else does. Just do the best you can. The world doesn't seem to care one way or the other."

"What other advantages are there to foreseeing the future? I keep getting the feeling that you're not telling me everything."

LeRoi thought for a second. Maybe now was the time to clarify this business about changing the past.

He reached over and positioned the napkin holder flat against the wall. He then moved the saltshaker to the middle of the table and the pepper shaker to the edge of the table. He now had a straight line consisting of three points, napkin holder, saltshaker and pepper.

Jason gave him a strange look. "Are you about to show me another trick?"

"No, this is no trick, this is a timeline. In fact, this is your timeline, a graph of your life."

Jason looked at the props on the table.

"This is my life?"

"Yes, think of the napkin holder as your past, your birth would be a good date to put on the napkins. The salt is the present. Think of yourself as being the salt of the earth. The pepper is your future."

"What's the date on the pepper shaker? If that's the end point, my death, what's the date?"

"It says in the Bible, 'No one knows the time or the place.", said LeRoi.

"Wait a minute, didn't Jesus know, didn't Martin Luther King know? Haven't a lot of people known?", asked Jason.

"They may have known, but let's just say there are exceptions. I certainly don't know what date to put on the pepper shaker." There he goes off on another tangent LeRoi thought.

He brought him back to the subject at hand. He reached out and moved the salt a fraction of an inch toward the pepper shaker

"What are you doing?" Jason was watching him closely. He didn't want to be tricked into buying another round of drinks.

"This is a Real Timeline, and the present is dynamic. Every minute which ticks off the clock moves the saltshaker a minute closer to some unknown pepper shaker."

Jason felt a slight chill as he imagined the minutes ticking off his life. It all seemed so finite. He suddenly realized just how mortal he was. LeRoi reached up and moved the salt another fraction of an inch towards the pepper.

"Would you quit doing that?"

"You may as well face it, we're all on a timeline, moving towards the end. Once a person accepts that fact, they begin to appreciate all the blessings around them.

"There is really no time in this world for hate and misery; too much out there to love and enjoy. So much to do and see and so little time remaining. But I'm getting away from my point again."

"You are here at the saltshaker," LeRoi said. "Let's say you can project yourself into the future, let's say somewhere up around the pepper shaker." but your body is still back there at the saltshaker which represents the present or Real Time."

"That's correct, so your mind is close to the pepper shaker, and you see something you don't like."

"You mean like a plane about to crash, with me in it?"

"Yes, that's a good example. The saltshaker becomes the past?"

Jason had to think a minute. The saltshaker which represents Real Time would be in the past."

"So, the next thing to do is to go back to the saltshaker, the present, and modify the events so you don't end up in the airplane."

Jason caught on immediately. "What you're saying is, if you can see the future, you can change it by changing the past. But you said earlier that you can't change the past."

"Well, that's generally true, but when you look at it from the vantage point of the future, you can change the past. If you can get ahead of it, it can be done. But remember you're not really changing the past; you're changing the present. It is only the past in reference to a point in the future. After that moment in real time goes by, it can no longer be changed."

Jason had that look on his face which showed he was almost to the point of understanding.

LeRoi lay his crippled hand flat on the table. "You see the spot where the finger used to be?"

Jason looked at it. It was just a stub. A hash mark scar with a few purplish lines radiating away from it.

"If there was a way to change the real time past, I could go back to my childhood, modify the events of June 6th, twenty years ago and we could both sit here and watch my finger reappear."

Not that he hadn't tried. He had relived the mistake many times in his mind. His finger had never reappeared.

"What happened to your hand?"

LeRoi hadn't thought about it for some time. "It's not much of a story."

He had been at his Uncle's farm in southern Illinois. The day of memory was one of those hot, sticky 'Dog Days' of August. He'd gotten up with the sun, pulled on his overalls, shirt and work boots. He was going to the fields with the men. Acres and acres of golden hay to bale. Men's work, and he wanted so much to be a man. The hot hazy morning of twenty years ago unfolded for him.

His Uncle Frank was up front on the tractor and his cousins, all older and bigger, were on the back of the wooden farm wagon. His cousin Carl was tall and muscular. Only two years older than LeRoi but with the physical tasks of the farm, and the three big farm meals a day, he was about twice LeRoi's weight.

"You think you're big enough to bale hay? Why, I reckon you don't even weigh as much as one of those bales."

Carl was wrong, the bales weighed sixty pounds apiece and LeRoi weighed at least ninety. Carl went on to explain how to handle the weight.

The first bale came slowly up the chute. Carl grabbed it, swung it up in the air and towards the back of the wagon. The bale settled into its position, plumb and square with the back of the old wooden frame.

LeRoi was ready for the next one. He grabbed it, gave it a mighty swing, but didn't let go of it in time. The bale slid to one side, got too far of the edge, and fell off the side of the wagon, pulling young LeRoi with it.

He remembered it all as if it had just happened. The embarrass-ment of making a fool of himself in front of the people he looked up to. He scrambled to his feet and dragged the bale along as he caught up with the slow-moving hay wagon.

Carl jumped down, grabbed the bale and helped him throw it back on the wagon, vaulted up on the flat bed and helped LeRoi back to his work position. "You're s'posed to throw the bale, not let the bale throw you," Carl said good naturedly.

LeRoi caught on quickly. He'd worked harder that day then he'd ever worked in his life. By day's end he was badly fatigued, but much proud. His hands and arms were scratched. and cut from the rough straw. Every part of his being was exhausted. It seemed as if every drop of moisture had been drawn from him. The dry chaff clung to his skin and dusted his jet-black hair to a speckled white. He felt exhausted, but whole. A man's work and he done it all day long. Other than one stack falling off the tail of the wagon because he'd forgotten to cross stack them, he'd done a good day's work.

As the tractor and hay wagon pulled into the barn yard, LeRoi was basking in his own glow. Like a proud warrior coming home from battle he was puffed up and arrogant.

He placed his hand on the floor of the wagon and vaulted to the ground. He thought nothing of the ring on his right hand. He subconsciously felt the head of the ten-penny nail sticking slightly above the bed of the hay wagon with the palm of his hand. It was only an eighth inch above the bed, a fraction above the flush line. There was no need for his mind to send up a warning flag. No reason to think anything about one old nail a fraction of an inch out of position.

The old hay wagon, loaded with hay and weighing in at twenty thousand pounds, was rolling slowly, being pulled in one direction. LeRoi Varitia, weighing in at nighty pounds and traveling in the opposite direction was being pulled by the force of gravity and the slight push he'd given himself. He was moving at about twelve miles per hour in the opposite direction.

The only obstacle restricting the two objects from going their separate ways was a small dime store ring of steel and the head of an old ten penny-nail.

Uncle Frank never felt the slight jerk of the wagon, though later he would claim he felt a dart of pain shoot up his arm about the time it happened. Sympathy pain was what he guessed it was. Sympathy for the ring finger LeRoi left lying on the back of the old hay wagon.

Carl watched the whole affair. He would remember the hard jerk that straightened LeRoi's arm and threatened to dislocate LeRoi's shoulder. It happened so fast. He looked to the bottom of the wagon to see what LeRoi had caught his hand on, but what he saw was the upper two joints of a small pink finger. There was a little splattering of blood where the finger lay. It all looked so gruesomely out of place. A Finger laying on the flat bed of an old, worn wagon.

Carl turned a little white. He felt nausea for a second, and then regained his composure. He remembered taking off his hat and throwing it down over the finger. He jumped down off the wagon and walked back to LeRoi.

LeRoi had gotten to his knees. He was holding his crippled hand. He was holding tight to the most painful appendage that could ever exist, but what was confusing was, he was holding onto something that wasn't even there. How could something hurt so bad that didn't even exist?

By now Uncle Frank had stopped the wagon and walked back to the scene of the accident. He quickly evaluated the situation and pulled his red bandana out of his back pocket and tore off a thin strip of cotton. He tied off the stub to stop the bleeding. LeRoi was close to passing out from the pain but he managed to stay on his feet.

To LeRoi it would always be as vivid as if it had just happened. He could travel back to that instant and review it.

He had done that many times. He would review it right to the point where he put his hand down on the head of the nail, only in his mind he would change it. He would put his hand down a bit to the left or the right.

He would jump off the wagon and imagine that the nail would not grab him by the finger. He would return to the present and look at his hand and hope to see his finger reappear, but it never worked. There would only be the stub where his finger used to be. He remembered the words of his Aunt Ruth as she sewed the flap of lose skin over the wound. "That finger's gone now. You got nine more to take care of. No need worrying about the one you lost." But, he had

worried about it. He'd tried a hundred times to get it back.

LeRoi looked across the table at Jason. "I lost that finger on Real Time. There is no way to get it back."

Jason thought about the story. "If you could have seen the event happen, like maybe in a dream the night before, you could have avoided the accident. Had you raced forward in Mind Time and raced back to the present, which was actually the past compared to where you would have moved forward from this... was becoming too complicated. I guess I see your point, but it gets confusing so fast."

"Just keep the basics in mind. If you get confused, you're probably going down the wrong road.", said LeRoi.

"Ashtrays, Motherfuckers!" Frank called out over the noise of the bar. The blatant vulgarity brought the lesson to an end.

"See you next Friday Jason."

"Yeah, next Friday." Somehow weeks no longer seemed like weeks to Jason. He didn't know what they seemed like.

PART TWO

Chapter 13

The weekend went well for Jason. He'd gone home to his mother's. Instead of drinking himself into oblivion, he'd spent the early morning hours writing a letter to his son.

Saturday was spent weeding his Mother's garden and Sunday he actually went to church. Not that he received any revelation or insight. It seemed he was much more interested in a brunette who sat two pews over.

Jason woke up early Monday before the alarm. He felt renewed. Could a little conversation and some garden work renew you? He didn't know. I'll need to ask LeRoi about that next week, he thought.

'Up in the morning and off to school,'

'Always remember the Golden Rule.'

The old song was rattling around in his head as he drove to work with a new dedication. "I'll not let the assholes get to me anymore," he said to the window in the old van. Life ain't so hard, he thought as he pulled up to the rail on the blacktop parking lot. He punched his card and headed

down the long aisle. How many days had he made this journey? How many more till he reached the pepper shaker? The lessons in time travel seemed to grow dim as he approached the real world of hard work and sweat.

Jason changed from his blue jeans to his thermally insulated underwear. It was going to be another scorcher. Long cotton underwear kept the radiant heat off his legs and gave him an extra layer of protection in case he nuzzled a little too close to the fast-moving, hot copper bars. He pulled on his long blue work pants and got into his greasy work boots. I'm about ready for a new pair, he thought. Had he been able to read the future, he would have known he'd never finish wearing out this pair.

The summer heat had been getting on everyone's nerves. There was the usual amount of practical jokes and with the heat as intense as it was, some of the frivolity had turned sour.

Last Friday, Jason had been in the shower, washing his hair, when someone had reached over and shut the hot water off. The shocking difference of going from warm to cold should have been a laughing matter, but when he'd opened his eyes and seen Billy Joe standing there, he'd almost taken a swing at him. The anger came and went in an instant, but both men knew it was there. With Billy Joe there was always a tension...a distrust maybe even a slight hatred.

Jason hadn't bothered anyone since the fuse and shut-off business on the line. He'd taken whatever the mill men could give and gone on about his chores. Most days his only interest was improving his skills, finishing his work and getting out of the heat.

The heat was the real enemy. It drained your strength. It sucked every ounce of water from you it could. If you made a mistake in this weather and had to cut a bar or roll hot scrap, it was almost unbearable. Today would be no exception.

Big John was in the cafeteria before Jason got there. He was leaning against the old chest-high space heater eating from his large brown paper sack which seemed always to be full of food. He finished a sandwich, threw the waxed paper away and pulled a fresh box of corn starch from the bag. He unrolled a big red bandana and laid it on the heater. "Gonna be another hot one today, Patch. You gonna be perfect?"

"Just like every day, I never make a mistake." Both men laughed as John opened the new yellow box of cornstarch and poured a pile of the white fluffy powder onto the bandana. He rolled it up and tied it around his neck. It was a ritual all the men practiced on the hottest days. The cornstarch-laden bandana kept the heat off their necks and stopped the rash which accompanied this type of weather.

John unbuckled his belt and poured a heaping helping down the front of his underwear. The excess white powder puffed out each pants leg, dusting his black work shoes with speckles of pure white. John hitched up his pants, unbuttoned his shirt and patted the powder under both arms. The pure white of the powder contrasted sharply against the deep ebony of the old man. He handed the box to Jason and Jason went through the same ritual.

The cornstarch was obtained free from the storeroom and both men used copious amounts. You could never get too much starch. That and the rock alum each man kept in their pockets kept their balls from galling against the insides of their legs. To be galled only added to the misery, making the long hot days unbearable.

They were both as ready and as dry as they could be, and they headed for the Hill. Each knew that the radiant energy emitting from the hot metal would soon create a river of sweat that would engulf them both. Soon their shirts and pants would be wringing wet. Efforts to stay cool would give way to the drudgery of the mill. Each would at first attack the work, be engulfed by it, and in the last hour attempt to endure it. Jason's body was adjusting to the severe conditions. Never again would the outdoor temperature feel hot to him. The man-made heat of the factory was conditioning him to withstand and endure.

There were four men on the Hill today. They were rolling five-sixteenth inch diameter copper rod and the smaller diameter required an extra sticker. That was no problem as everyone was in today, including both foremen.

With Old Al helping on the line, the stickers would get double breaks because of the heat. John was at his usual position on number one, Jason was on number two, T-Bone was at three, and Billy Joe was at four. Each man was a pro, and the day should have gone smoothly.

John called for the first bar. The long flat table leading to the first stand glistened with fresh oil. The heat waves radiating from the flat steel gave off a visual watery effect. The first bar nosed its way through to the first station and the day began like many others. John caught the first few bars and blew two shorts on the mill whistle.

The old monster of a looping mill rumbled up to full speed. The day went well for the first three bars, a total of twenty-four seconds. The fourth bar hit the entrance guide with a normal thud but as it came out the exit hole it brought the exit guide with it. The guide blew out of its holder. Had it not been for the hydraulic water line, it would have hit John full in the face. As it was, all it did was cobble. The exit guide knocked John's tongs out of his hands and showered him with hot dirty mill water.

Jason blew the stop whistle before anyone else had seen the error and the hot bars stopped coming with a minimum of four bars scrap.

"You OK, John?"

"That MoFucker 'bout took my face with it Who the fuck set that guide?" The question didn't deserve an answer and John didn't expect one. It didn't matter who set the guide, no matter how many precautions were taken, in this heat, hold-down bolts would come loose, and schemes would sometimes slip out of position. Both foremen were there to help get the mill rolling again, but John wanted to yell at someone. He had stayed dry for only twenty-four seconds and there was a long hot day ahead. He held his temper. He knew he would need all his energy to just get through the day. He wiped off his face as best he could, and the day continued.

The mill water had a certain rancid smell and John wished he were in the shower getting ready to go home. Too early to start thinking about the shower. It would only make the day go slower.

Halfway into the day one of the tubes fouled at the finish mill and Billy Joe ended up with a pit full of scrap.

Everyone should have pitched in and helped, but John and Jason went down for a cold drink of water and were slow getting back. That in itself set Billy Joe to brooding.

Next Jason got caught in a daydream and missed a bar. It should have been an easy bar to cut, but his foot slipped on some mill grease and he cracked his shin on the deadman. Had it not been for his long underwear, he would have had a nasty gash. As it was, he just did a little dance for the men as he felt the blood soaking into his long underwear on his right leg. Everyone went for water and Jason was left to pull his own scrap.

Next a guide blew on the back of T-Bone's mill and the overhead crane had to be called up to get the hot copper spaghetti off the coilers.

These were not major problems, just a combination of small mistakes. Finally, it was the last break before the end of the day. Jason and John were both in the cafeteria at the same time.

John smelled of mill water and wasn't the least bit happy about it. He pulled a big red apple out of his feed bag and sat it right in front of Jason. Jason knew better than to ask for the apple. It looked exceptionally delicious. John pulled out a sandwich and began to eat. Pretty impolite, Jason thought to himself. It was if John was flaunting the last of his food in front of him.

"John Curry call two-five-O."

John hardly ever got phone calls while working. The message came over the

intercom. It was part of the company paging system. The system allowed them to go to any of the page phones hanging at various places around the plant and receive their message. John went out the cafeteria door leaving the big red apple sitting in front of Jason.

Only a few more minutes and Jason would be back on the Hill. He dreaded the thought. Thirty more minutes of hell and he would be in the shower. Another day almost complete.

The apple sat there asking him to take a bite. He smiled to himself, imagining John coming back and seeing a big bite out of the apple. The old African would go into a rage. Jason knew better. No way could he bring himself to touch the apple.

Billy Joe walked into the cafeteria. "You be due back on the Hill in two minutes, boy. Them hot bars be calling your name. You ain't got time to eat that apple, do you?"

Jason hesitated. What he should have said was. "It's not my apple." Or "That apple doesn't belong to me." Or "That apple belongs to Big John and no one fucks with John's food."

The hesitation caught him. It put him in a trap he promised himself he wouldn't get caught in. He realized his mind was racing forward. He could see the big man walking into the cafeteria and catching Billy Joe eating his apple. All this he saw, and it only took a fraction of a second in Real Time. Billy Joe was standing waiting for an answer. His eyes were demanding an answer.

"No, I don't have time to eat it," Jason replied.

"Mind if I eat it?"

Jason knew better than to do what he did next. Al had warned him about practical jokes, how they sometimes escalate into and beyond what a person intended, but it was too late. He hadn't done anything on purpose. He hadn't wanted to set Billy Joe up, at least not consciously. Jason reached up and lightly rubbed the burn scar over his right ear. It was one of those habits which develops and then doesn't easily go away. "No," he said contemplatively, "I ain't gonna have time to eat that apple." Jason turned and headed for the Hill. Ten steps from the door he almost turned and went back, but he continued. He went up to the number two station and relieved the foreman, who in turn relieved T-Bone. Jason caught the next bar and glanced over the mill, to the cafeteria. He could see Billy Joe leaning against the space heater, chomping into the last few bites of one insignificant apple.

He watched as John hung up the page phone and walked towards the cafeteria. He caught another bar, stuck it and looked back over the mill just as John walked through the door. Jason's heart was clicking up to a faster pace. Suddenly he realized that Real Time seemed to be slowing.

There was no way he could hear a conversation from his vantage point, but he wasn't sure if any words were even spoken.

Billy Joe was right in the middle of a bite when John came through the door. The eyes of the big man stopped Billy Joe before he finished the bite.

The cold rage which emitted from those eyes would have stopped a charging elephant and within an instant, Billy Joe realized he was eating the wrong apple. He tried to get his mouth away to explain he hadn't meant to eat it. He wanted to put his words on fast forward, anything to keep the fury away from him.

He was too slow. Even John didn't understand the reasoning, for there was no reasoning. just rage and action.

Jason caught another bar, stuck it and looked back down to the cafeteria.

old Al, on the line saw Jason watching the cafeteria and looked over to see what was going on. He looked in time to see Billy Joe's head about a foot higher than it should have been, pinned against the coffee machine. John had both hands-on Billy Joe's neck and Billy Joe was flailing at the big man's arm's, to no avail. Al headed for the cafeteria and got there before Billy Joe passed out.

"Let him go, John!" The old man's voice wasn't loud, but it was demanding. "John let him go. John let him go."

Billy Joe slid down the face of the coffee machine and slumped to one knee. He quickly got his wind and tried to set the record straight.

"It was Patch told me to eat the fucking apple. I wouldn't touch your food. It was Patch, I tell you."

John didn't believe him. "Patch wouldn't do any such thing. Yous a liar. You stole my apple."

"It was him I tell you." Billy Joe headed out the door. Jason watched as the now enraged hillbilly headed for the Hill. Billy Joe knew he'd been had, but it didn't make any difference. He realized, as he thought back through the conversation, Patch hadn't exactly said it was his apple but that didn't make any difference either. He'd taken some abuse and now he was going to give some back.

Jason watched him come. There was no doubt what his intent was. Now Jason's mind was really racing. The bars seemed to be coming in slow motion. Billy Joe turned at the coilers and headed for the catwalk just below the Hill. The catwalk spanned the long slippery slopes that fed down and away from the stickers.

Billy Joe was on the catwalk now and it was obvious that the war was on. One hillbilly against one smart-assed young sticker. Jason watched him come. He thought he might be able to take him on neutral ground, but he wasn't sure. What he for sure didn't want to do was fight amongst the hot running copper.

What to do? Billy Joe came closer, now he was almost to Jason's aisle. He would be turning to come up after him. They all saw him coming. They all saw the anger. None but Jason knew why.

The copper bar was also coming. It hit the entrance mill with the same dull thud as always. The foreman at the first station caught it and passed it to Jason. Jason went down for the catch, but realized a solution lay right within his reach.

Instead of making the catch, he hesitated a fraction of a second, a moment's delay. He reached down below the bar and used the top jaw of his tongs for a guide. He turned his wrist ever so slightly and the copper nose lifted out of the track and stayed almost parallel with Jason's feet. Another slight adjustment and the bar turned toward Billy Joe. Now it ran steady and true.

At first Billy Joe didn't realize what was going on. Of course, no one understood it. It had never been done before.

Old Al had followed Billy Joe out of the cafeteria and had kept up as best he could. He knew Billy was going after Jason and he was coming around the coilers when he saw something he'd never even imagined. A long piece of red-hot copper, the nose only a few feet from the hillbilly's face.

The front pulpit operator saw the awesome sight at the same time as Ken Morehead. Both came out of their chairs. The pulpit operator was reaching for the mill controls. Ken was headed for the door. Both men were much too slow to do anything about what was about to happen.

The Foreman who was relieving Billy Joe was within reach of the emergency stop button. He also saw the cherry-red

guided missile. His heart nearly stopped in his chest and

his body froze. He wanted to reach out and slap the button, but his reflexes were shocked into submission.

There was only one man in the world who could tame the red monster, and he was the one who had turned it lose.

Jason waited. It at first reminded him of his early days playing with a hose in the back yard. He remembered putting his thumb over the end of the hose and forcing a stream of water at his older brothers. He could squirt anyone he wanted. A slight movement of his thumb in concert with his wrist and the stream of water would go anywhere he commanded. Once the water was out of the hose you couldn't call it back.

Billy Joe realized what was happening, but he was subject to the physical laws of gravity and reflex. It hadn't been a good day for Billy Joe, or for anyone on the Hill. The mill men would later remember it as the day in which everything that could go wrong, did.

First, there had been all the production problems. The blown guides, the blocked tube and the mess behind the mill. Then Billy Joe was almost choked to death by a raging madman. Now this was something that made the other problems look like child's play.

Billy Joe was confronted with the end of the world. The end of his world lay two feet away, traveling at 53 feet per second and sparkling at 1500 degrees Fahrenheit.

Billy Joe was like a baseball player trying to get out of the way of a fastball that was headed right for his ear. His legs went out from under him, like a batter's legs will disappear trying to duck the ball. It was to no avail. He knew he was a dead man.

He remembered seeing the nose of a bar hitting a bucket of grease once. The bar had pierced the metal bucket on one side, set the grease on fire in the middle and shot out the other side without even upsetting the bucket. He realized it would be his brain bursting into flame. His legs went limp, and gravity seemed to be his only hope for survival. Gravity wouldn't help in this case. Both he and the bar were at the same height. Both were suspended in midair. One hillbilly and one very hot, two-hundred-forty-pound piece of copper rod. Billy Joe's eyes grew very large as the bar came on, now only twelve inches from his face.

Jason had aimed his weapon. He'd pointed it as if it were a gun. A water hose wouldn't be a fair comparison. A stream of water wasn't nearly as deadly as this. A gun at close range would not have been as lethal. A bullet wouldn't first explode your brain and then set it on fire. To Billy Joe, it all happened so fast. At first, just a hard day in the rolling mill. Within a few seconds the entire world went on hold, encapsulated within a moment which seemed to transcend time and space.

One moment you're here the next moment you've stepped over to the other side. He saw glimpses of his childhood, his adolescence, and his younger days, growing up in the bluegrass country of Tennessee. Warm sunny days on mountain slopes, fields of wildflowers, mountain streams so cool and crisp you'd never want of a cool drink. How silly to be angry over insignificant problems. How foolish to waste your life in the bowels of a rolling mill. All this he saw and more as Real Time continued one millisecond at a time.

Jason's mind was racing also. It began as a thought, a solution to a problem, but the solution immediately became more severe than the original problem. He had felt trapped. He was on the Hill and the Hill was the last place he wanted a confrontation with an angry hillbilly.

The solution had been spur of the moment. Had he had time to think about it, he would surely have rejected it. To begin with no one had ever tried to use a hot piece of copper as a defense tool. Who would've thought it could be done? Old Al hadn't even tried this trick. It was an experiment. An experiment that had worked better than anyone could imagine. He had tweaked his tongs at the perfect angle to guide the hot bar up and out of its normal position. He turned his wrist at just the right height. He had turned the hot snake loose and it ran straight and true towards its target.

He wished for it to come back. All he'd wanted to do was stop Billy Joe at the catwalk. That had been accomplished. Now he wanted to end the deadly game of "fuck you very much." The nose was a long way out.

The longer the nose is away from the sticker the more momentum. Had he been at the first station where John worked there could be no hope. Even a man as big as John couldn't pull against both the static load of the bar and the dynamic load of momentum.

Jason rolled both his wrists and grabbed on to the fast running bar. It was like grabbing the bumper of a rolling car. He set his right foot against the bottom of the steel deadman and felt his shin dig into the top lip. The cut on his leg from the previous mistake split further and soaked his lower leg with blood. He felt the pull in his shoulders. They wanted to come out of their joints.

Had his hands been a bit more damp the tongs would have slipped and been carried with the bar. Had his back not been so strong from a thousand catches, it would have given way to the tremendous force. But as it was, everything held in place, including the nose of the bar. It came to a stop one inch from one middle-aged man suspended in mid-air, one inch from a murder charge for Jason.

As Billy Joe hit the floor, the bar dropped to the catwalk rail and as Jason gained complete control and dragged it back across the deadman, the nose slipped to the greasy hot steel and slithered away like an angry red snake.

Jason tapped the hot copper to the soft steel, raised his ax and struck hard to the bar. It cut clean. He double caught, turned and slammed the new cut nose to the waiting rolls. As the rolls grabbed, he heard the next bar being delivered, turned and caught it long. He worked his tongs up the bar and completed the stick. It seemed an eternity before the next bar got there.

The superintendent was out the door and on the overhead catwalk. He couldn't believe what he'd just seen. He was prepared to move down the steps two at a time, but he stopped where he was, not knowing what to do.

The old mill was humming along as if nothing had happened. He was sure someone would need to be disciplined but he wasn't sure what crime had been committed.

Al had just turned the corner at the coilers when he'd seen the impossible. First an incredibly foolish act of skill and control, and then, when he was sure he was about to witness a death, everything returned to normal. The words he had spoken to Ken came back to haunt him. "What can he do that we haven't already seen?" is what Al had asked. Now they knew... The day was about over but Al knew it would never be forgotten.

The front pulpit operator had his hands still resting on the emergency stop controls. He realized just how little control he really had. He had always imagined that if a life threating situation occurred, he would be able to assist the stickers and get control of the problem. This one had come and gone and there was nothing he could have done. He stood there at the controls. He felt he should do something but there was nothing to do. Everything was back to normal.

Everything was not back to normal. Billy Joe was getting slowly to his feet. He knew he was all right. He had ten minutes left in his break, but he didn't want to go back to the cafeteria. He walked up the aisle to his station and stepped in. The foreman handed him his tongs and stepped out.

"You got ten minutes left afore you're due up here."

"Give the time to someone else, I'll be finishing out the day now. You might want to roll that scrap before it causes a cobble."

The foreman pulled on a pair of leather gloves, slapped some mill grease to them and walked to the aisle where the hot copper lay half in and half out of Jason's pit.

John had followed Al out of the cafeteria and had come around the corner in time to see Billy Joe hit the deck. At first, he thought there had been an accident and Billy Joe was just dodging an errant piece of wild rod, but as he watched Jason grab and cut the bar, he realized it had not been an accident. He didn't like Billy Joe, and Billy didn't really like him, but they'd always tolerated each other. This deed that Patch had done was beyond the scope of dislike. This was a deed that should never have happened.

John felt bad about what he'd done. Felt sorry as soon as he'd realized that he had been set up, but it had happened so fast. Now he was faced with having to come in everyday and work with a man he'd almost choked to death.

Al began walking back to the showers. It was a little early to be going but he often left early on hot days.

Ken came down the stairs and waited for the foreman to finish rolling the scrap. He motioned for him to come over.

"Get a drink of water or whatever you need and then finish up the day on number two. Tell Patch to hit the showers. I'm not sending him home. I just want him off the Hill until I can figure out what the hell happened here."

The foreman headed for the cafeteria and John walked up and relieved the number two foreman at the first station. He looked over at Jason. "You tell Billy Joe to eat my apple?"

Jason didn't answer. He was wired and his mind was still racing at about a thousand miles per second. The bars felt light as a feather compared to the one, he'd pulled back. Everything in his body should be hurting, but he felt bulletproof. He was pumped.

He didn't think he could even talk in a regular manner, so he kept his peace. His silence would have to be answer enough. The lead foreman walked up to the second station and stepped in. "Ken says hit the showers. I'll cover the last fifteen minutes. You're done for today, Patch."

Any other day, Jason would gladly have gone out early, but he was still mad. Angry with the heat, the heavy work and, most of all, angry at all he'd endured. Just when he thought he'd found a home back in the rolling mill, he'd blown it.

"It's not my fault!" He was saying to himself. I did the best I could. But his rational mind was screaming at him that he'd fucked up big time, that he'd opened some Pandora's Box and there was no going back.

That made him angrier, like he'd lost something he'd tried very hard not to lose. Anger and frustration welled up inside him and he felt like crying. Always, instead of releasing tears, he released fury.

"Give the fucking free time to someone else. I don't need it."

"Ken says you hit the showers. You'll just be making things worse if you refuse a direct order."

The next bar hit, and the foreman reached down and caught it. Jason angrily stepped out and headed for the aisle leading to the shower.

He actually bounced when he walked. His legs and body were wound tight. He felt like he was going to explode. He walked past an old coil of scrap and swung his tongs down

hard on the soft copper. The tongs sunk in with a thud.

"Fuck you, Mr. Copper Coil!"

He continued on past the line and swung at the old empty relay box which had been used to make a fool of him. He swung the old tongs at the box and the steel on steel echoed down the line. "Fuck you, Mr. Relay Box!" He walked towards the time clock and took a swing at it. It dinged just as if someone had placed a timecard in it. A flake of paint flew off the old green box and floated to the greasy aisle. "Fuck you Mr. Time Clock!"

"I whipped your ass, copper mill." He said to no one in particular. "I whipped your ass yesterday, I whipped your ass today, and in twenty years I'll still be whipping your ass," he yelled as he walked through the shower room door.

Old Al was sitting on the locker room bench as Jason walked by. He didn't look up. He sat there naked unmoving, ready to go to the shower. Sitting motionless, as if the effort would be too much for his old bones. His head was bowed from the years gone by. His thin grey hair lay lifelessly against the slightly translucent skin on his balding head.

Jason sat down and pulled off his clothes. All pumped up and nowhere to aim it. He felt strange and almost giddy as he hung his clothes in his locker.

As he headed for the shower, he noticed old Al sitting there motionless. His skin hung loosely from the bony old frame of a body. His scrotum cascaded over the edge of the bench like a wrinkled waterfall emerging from a kinky mass of grey-black hair. There was something different about him now. He hadn't moved but he didn't seem to be the same person that Jason had seen moments earlier.

As Jason walked by, he stopped and looked back to determine what was different. What he saw caused his heart to stop. It wasn't Al at all. It was someone else. This new man was sitting at Al's locker. He was Al a minute ago but now he was someone different. Jason stopped and looked harder. It was like he was looking through fogged glasses. The image was there, and it was familiar, but he couldn't figure out who it was or how this strange new man had replaced Al without Jason noticing the switch. Hadn't he seen Al sitting there when he'd walked in? Maybe he'd thought it was Al and it had been this other man all along.

There was something familiar about this old man who sat before him with his bag hanging lower than his old, wrinkled cock. This man had a burn scar on his right ear. Not too unusual, Jason thought, as he reached up and felt his own ear. The scar was close to being exactly the same as the old man's. In fact, it was exactly the same. Jason felt the bile start to rise in his throat as he realized he was looking at himself, twenty years into the future. What had he just shouted as he walked through the door? "I'll kick your ass twenty years from now!" Well, there he sat, twenty years older, and it didn't appear he was kicking anyone's ass. In fact, it appeared he had taken on Father Time and was losing...losing badly.

The shock of seeing himself in such disarray and defeat made him physically sick. He stumbled towards the toilets and quickly lost what little lunch he had left.

He weakly stood from the stool and moved towards the bench, but it was no longer himself sitting there, it was old Al. No burn scar. No vision of his future self, just an old, white-haired man.

Al looked up. "You OK Patch?"

"I'm OK, just too much heat. Too much of this old mill. They just looked at each other for a long second and then Jason moved off to the shower. He wanted to wash this day off, but it didn't work. The sweat and dirt and copper washed away, but this day was stuck forever. It was part of the Real Time past which made up everything he was. The Real Time events which would never go away. It would need to be accepted before he could go on with his life.

It was like LeRoi losing a finger. The difference was, this wouldn't be a physical scar, it would be a mental scar.

He had made a trap for himself, and escaped with only a bad memory, but it could have been much worse. He had the blood on his hands from what might have been. He'd murdered a trust which could not be brought back to life. He'd pushed himself up to the edge of an event he'd wanted no part of. Now he was stuck with mental scars that would be with him forever.

Jason pulled on his clothes and headed for the time clock as everyone else headed for the showers. He could hear the old mill winding down. He could hear all the sounds of the day ending, but he somehow felt separated from it. He was no longer a part of the system. He walked past the stickers as he walked towards the time clock at the far end of the building. There was no 'See you later, Patch,' or 'Have a good day Patch'. It was all silence. It was not something planned, it was just that no one had anything to say. It was a very lonely, long walk to the parking lot.

As Jason got into the car and headed for the gate, he got an old feeling he used to experience about this time of year. He didn't know if it occurred due to a fresh northern breeze or if it was something in the color of the trees and the grass, but it was remotely familiar.

Instead of turning south to go home, he turned north to the college campus at the outskirts of town. He felt a desire to go to school. The thought made him laugh out loud. Thinking himself a student again was rich, he'd never really liked school, and had it not been for baseball, he probably would not have made it as far as he had.

He had a vast emptiness within, and he was seeking something to fill the void. Maybe he could walk around the campus and try to rediscover that old 'first day of school' feeling.

He pulled his van into the lane marked college and followed the green signs marking the way to the 'Visitors Parking' section.

He was afraid to get out. The parking meter stood like a sentinel guarding the lot from foreign invaders and foreign was the best way to describe the way he felt. He'd never

been on campus before and the illusion of what it would be like frightened him. He realized he was most afraid of being around people he felt were much smarter than himself. He was beginning to feel the exhaustion of the day's work and mistakes. He thought about backing up and going home but something told him he had no home.

"What do I really have?" He thought. His marriage was gone, his savings were all gone, and he was twenty-seven years old and living with his mother.

The only thing he really had was his work and now that was also gone. Oh, he knew he could go back, but it wouldn't be the same. The men would all be mad at him for a time and the way he would be treated would never be like before, but he could either endure it or bid back to another area. The reason he knew he was done was the vision he'd seen of himself.

The words of LeRoi came back to him. "You can change the past, if you can see the future." Oh, there was no way to change this day in Real Time. No way to change the things he had done, but there was a way to change tomorrow, and tomorrow was the past viewed from the vantage point of the future.

He realized he was at a turning point. Most of the time he had only been able to look backwards and see the turning points. Most of the time he could only see the milestones in his life by looking back and saying, "If only I had known, I would have done things differently, but it's too late now to change anything."

Well, it wasn't too late. He had a choice. He could either get out of the van and find a way to get back in school, or he could back this old van away from the parking meter and continue his timeline in the exact way it was headed.

The vision of himself sitting at the old wooden bench, wrinkled and defeated, surfaced in his mind. It was so clear and so shocking he shivered. The future was mapped out for him if he continued in the rolling mill. If he wanted change, if he wanted to begin a new timeline, all he needed to do was get out of the van, put some money in the meter and follow his new dream.

That minute little action. That insignificant effort would place him on a new timeline. He reached into his pocket and realized he was broke. Not even a quarter for the parking meter.

He leaned back and shut his eyes in exasperation. Where are you now, Jason Strider? What in the world are you doing here? What would LeRoi Varita say if he could see you?

Somewhere in the deep recesses of his imagination he saw a four-fingered, black-haired traveler canoeing down a river. He was a long way gone. Gone somewhere south. "Wait for me," he wanted to cry out. "You're not done with me yet. Didn't you tell me you'd teach me to travel in time? Didn't you tell me we could build a time machine together? Didn't you tell me you could teach me to time travel in just four lessons?" His mind did a quick count of the lessons. "Don't leave me here all alone!"

Jason could see a watery wake, glistening in the sunlight, a fine vertical line representing the tail end of a vanishing canoe. Once in a while the sun would catch the glimmer of a butterfly paddle as it flashed to stern and disappeared, dipping again into the green clear water.

What about the lessons? What about our dreams of journeying together through time? ...Rather rude of him to just up and leave like that, Jason thought as he climbed out of the old van. He looked around to see if anyone was watching and then looked back to the parking meter. "Fuck You, Mr. Parking Meter," he said as he turned and walked towards a big red brick school building. Somehow, he almost expected to see an old ramp leading down to a coal-fired

furnace, but what he saw on the outside of the building was a large white sign which said, "Admissions"

Chapter 14

Jason walked through the double doors that sat to the left of the big admissions sign. He stepped into the building noticing the cleanliness of the square tiles that made a mosaic pattern on the floor. There was soft music coming from somewhere, and up ahead lay a bustle of activity.

An information desk was situated at the confluence of three different aisles and opened to a group of centrally located offices. Jason stopped at the information desk noticing how pretty and young the two girls seemed. He was cautious, scared, and a little embarrassed at feeling so out of place.

"Uh, I'm kind of new here. Where would I go to see about starting classes?"

One of the girls looked up. "Which school are you interested in?"

Jason's mind did a quick trace, looking for an answer. This was a regional campus, offering courses from Indiana and Purdue university. One of his brothers had gone to Purdue and that was the first school that came to mind.

"Purdue."

"Will you be enrolling in engineering?"

His mind searched for an answer. He had an uncle that had been an engineer. "Uh, yes."

She pointed to a group of offices down the hall. Start down there the secretary will assign you to a counselor and he'll get you started in the right direction. It's just past the financial aid sections.

Jason walked into the office of engineering and talked to a secretary. "Hi, I'm Jason Strider. The girl at the information desk sent me here to be assigned a counselor."

"You mean the woman at the information desk?"

Jason thought back to the pretty young girl he had just talked to. Was she a girl or a woman? Was he making some kind of blunder already? Would he be able to say anything and not feel out of place? The pause must have been enough of an answer because she continued.

"What area of engineering are you interested in?"

This was almost too much for Jason. He paused again, waiting for his mind to kick out and answer. It all seemed so foreign to him, but he'd gotten this far. He waited for a bit longer. His silence had often worked in the past. People would answer their own questions if you gave them enough time.

"Well? We have curriculums in Industrial, Mechanical, and Electrical Engineering. If you're interested in Civil Engineering, you'll need to go down the hall."

Jason had been in heavy industry for seven years and he answered without any more hesitation, "Industrial."

"Two-year program or four-year program?"

"Uh, four years."

She opened an appointment book and studied it briefly. "Your counselor will be W. Q. Worthy. He'll be free Tuesday at… What's better for you, morning or afternoon?"

Jason again waited. The pause had been working well and he just didn't seem to be able to respond. This was worse than any of the jobs in the factory. Here he couldn't seem to get any words out. He hoped he didn't look as scared as he felt. He finally felt he could say something but before he could answer she grew tired of waiting.

"Just a moment, maybe we could get lucky." She pushed down on the intercom button and spoke into the gray box on her desk. "Mr. Worthy? Did you just have a cancellation?" A favorable reply came back through the static. "I have a..." She looked up making a question out of her expression.

"Jason Strider," He answered, grateful that he could finally answer an implied question.

"I have a Jason Strider out here," She continued, "That's interested in Industrial Engineering. Could you possibly squeeze him in right now?" A pause and then another affirmative answer came over the intercom. She pointed at a door down the hall and Jason walked to the door and entered.

Warren Quincy Worthy sat behind a cluttered desk and looked up, "Sit down young man and we'll get started." The old white-haired professor pulled some forms from a basket on his desk. "What say we get right to it, after all, you're already a week late as far as registration and classes start next Monday. I assume you have your SAT scores with you?"

"Uh no, I haven't, uh, I don't, er, I... well, you see, I've been working in a rolling mill for some time now, and I was just really seeing about the possibility of coming back to school, but, uh, I really don't know if it's even possible, but I walked in and just kind of ended up here and I don't know if I can get in or not."

"I did take the SAT before leaving high school, but I don't have the results. I remember they told me at the time I wouldn't have any problem enrolling in college with the score I received but I don't remember what it was.

WQ leaned back in his chair and looked at the young man across from him. He liked what he saw. Jason was obviously older than most of the students that came to see him, older and somehow more humble. Mostly he saw rich kids, kids that already had all the answers. This one was hungry, hungry for help... Maybe even desperate for it.

"Where are you working?"

"I've been working at the copper rolling mill on the east side of town.

"That's where they roll copper bar-stock into rod, and then make wire out of it?" WQ asked.

"Yes, I'm a sticker. Some call us rollers or edgers or catchers. That's what I was today, but if I can find a way to get back into school, I believe I'll... well, I don't want to go back but..."

WQ's eyes brightened. He knew about stickers. He'd seen them once on a plant tour. He also knew how rare it was to get one in his office. The factories in Fort Wayne were mostly one-way streets for those that entered. He even had some statistics on those that began their working days there.

The longer they stayed the harder it was to escape. The big 'dollar per hour' figure usually caught kids in a trap that closed tighter and tighter as the years went by. A wife and few kids and a lot of debt and soon the hole was so deep and the chains so tight that only a Houdini could escape. Was this what he had sitting across the desk from him? A Houdini seeking help? It was so rare to have a chance to save a drowning man that he felt obligated to help, if possible.

He also realized that this was foreign ground for Jason. A wrong move or discouraging remark may push him back into the factory or in another wrong direction. WQ looked at his watch. He shouldn't be in his office this late. He looked at Jason. No need to explain how much pull it would take to get it done. No need to bother him with the details of closed classes and late test scores and high tuition. If all went well and this long shot had the courage to get this far, the least he could do was give the kid a big push.

"Tell you what Jason, tomorrow's Saturday. Be here at eight AM. You'll need to complete some ACH tests which will show us the proper English and Math classes to place you in. After you've completed those, I'll review them and make sure your placed at the correct starting point. Monday you'll begin classes. I'll start you out with twelve hours, that's four classes. I'll get you started in the class level that I think will be right for you and if you have any problems, I'll transfer you into the proper slot.

Stop in my office about nine Am Monday and I'll give you your schedule. I'll get you a tuition deferment until we can push you through financial aid or until you find another way to come up with the money. By next semester we'll know if you enjoy being a student or not. Any questions?"

Jason sat stunned. He had heard horror stories about getting enrolled in college, how if you were smart enough to get through admissions you were smart enough to graduate. Here he was, already through. Did that mean he could graduate? He felt lightheaded. WQ sat waiting for an answer.

Jason looked into the old professor's eyes, into a vast resource of understanding. Here was man who had just helped him. Jason was sure that this was not the usual treatment.

He knew that this was more than just a conference. "Why?" Was what he wanted to ask, "I don't deserve it," was what he wanted to say. "I don't know if I can," was his biggest worry, but he held his silence and just nodded.

WQ smiled. "Well, get out of here," the old professor said as gruffly as possible. Be in my office bright and early Monday and good luck on the tests tomorrow.

Warren Quincy Worthy leaned back and stared at the off-white ceiling. This was to be his last year of teaching. It seemed everyone was anxious for him to quit. His students often made jokes about him when they thought he couldn't hear. The other faculty members thought him outdated and slow. The Dean of engineering respected him but was anxious to streamline the department with younger more up to date professors. WQ closed his eyes. He thought back to the days when he was no older than Jason. Back to when his professors had been people like Schroeder and Hientzberg. They were of the class of Albert Einstein and even though he had never met Mr. Einstein himself, some of his teachers had. He drifted back to those war years where excitement was everywhere. He began to relive a past he'd not thought of for some time. When he opened his eyes again it was seven PM, and everyone was gone. He stiffly rose from his desk, gathered the papers he'd need to push his young student into school and slowly ambled out the door.

His gray, short, cropped hair lay close to his temples and the skin of his hands showed the dark age freckles of all the years he had accumulated but tonight he felt younger.

Tonight, he felt like the old days before the Project. Tonight, some of the guilt was lessoned, some of the misery and worry of all those innocent people weighed a little less on his conscience. Tonight, he might even sleep.

Chapter 15

LeRoi Verita leaned back against the spare canoe paddle he had wedged into the stern of the sleek sawyer canoe. He adjusted the boat cushion between his back and the paddle and relaxed into a position to best catch the morning rays of the sun. He steered his canoe into the center of the river and caught the increasing feeling of floating through space.

His backpack, bedroll, and tent, up front, kept the canoe counter balanced and on a level plane. He felt some sadness at leaving, but it was that time. The cool breezes of autumn were pushing him southward with the birds and the monarchs. He thought back to the bar where he'd had his last conversation with Jason. He'd already went over it in his mind many times. He was sure Jason hadn't even noticed what he himself, had said, even thought it was Jason that prompted it.

They had been reviewing the lessons in time travel and Jason had introduced one of his own observations and said, *"You can begin anew with a kiss. One kiss started me on a time journey that allowed me to relive a very enjoyable part of my past."*

Then LeRoi had jokingly said, "Sure, I'll look into it. I'll go to a library and get a book called, '*Kiss for A New Universe,*' or '*In the Beginning There Was A Kiss.*' Jason probably hadn't noticed but wasn't that the way many beginnings were made. Simple observations made in jest that turned into something much more.

LeRoi had already lost much sleep over the off-hand remark. He had already gone through every science text he could find and there was not even the slightest hint that anyone had ever come up with this particular thought before.

His only reference was still the old fairy tale of the frog and the princess. This of course was as bad as his revelations about time travel. When you were able to see the future you often couldn't reveal it because you'd make a fool of yourself. This was even more obtuse, according to this line of thought, life on earth was closely related to a fairy tale.

He laughed at himself. That revelation certainly wouldn't get you into the pictures with Fermi and Einstein. He reached in his top pocket and pulled out his note pad and went over his thought tree again. It was so simply beautiful that he liked it more every time he looked at it. He put his notepad away and reviewed it again in his mind.

Einstein had related energy to mass and according to the physicist Stephen Hawking, Einstein had thrown the speed of light squared to make the two sides of the equation balance. That was all there was to Einstein's most important observation. It was left up to others to figure out fission, fusion, and the many other various, accomplishments that went with that single revelation, and yet, LeRoi, had never seen the significance of what it meant to relate energy to mass, but now it came crashing in on him.

The sun long worshiped by many cultures as a God itself provided a tremendous amount of energy to the earth.

Considering the surface area of the earth as a giant ball, rotating as if it were on a spit before the enormous hydrogen fusion reactor of the sun, and considering all that mass from the sun being converted into energy every second, then it seemed obvious to LeRoi that the earth must be gaining a portion of that energy and converting it back to mass at whatever the conversion rate was, which could be figured using Einstein's theory, of course that meant that the Earth was gaining mass every minute.

The next step in the observation involved asking about the other planets. If the earth were gaining mass every minute, then wouldn't all the planets be gaining mass for they were also rotating, as if on a spit around the all glorious sun.

The answer was, the other planets were probably not gaining mass. The reason being that it took more than just energy to create mass. It took a mass converter, and that mass converter was life.

Wouldn't that be a giant joke on the people of earth. That life, maybe all life forms, were just energy to mass conversions units. But how do you create life?

Do you just take a bit of salt water, add capillary action, and throw in a little sunlight, and wallah! There is life. No, there is no life without love… and that was the connector. '*Kiss for a New Universe.*' Could a universe begin with a kiss? The energy from the sun was there and it was enormous, but energy itself was useless if it couldn't be converted to something.

And what about figuring the formula backwards through time? The amount of energy converted to mass each day would depend on the amount of life on this sphere called earth.

The more life, the more converting every minute of every day. With a little calculous you could calculate the mass of the earth backwards… and wouldn't that projections, 120 million years back or so, give a much smaller mass of mother earth and wouldn't that smaller mass mean less gravity, and wouldn't less gravity mean things would grow much larger… like maybe as large as a Rhamphorhynchids or a Brachiosaurus.

But much more important than that, if any of these things could be formulated and computer modeled, wouldn't the connection between love and life and physics and mathematics and chemistry and philosophy. LeRoi laughed again. How funny it all seemed and how worthless he knew these ideas were if they couldn't be pulled together.

Needing to make something of all these ideas was one of the reasons for leaving. He had a friend in Ocean Springs, Mississippi, not far from New Orleans, that was a chemical engineer. Another teaching job awaited him there and his friend would help him with the mathematics. The power plant where his friend worked had a good-sized computer that was accessible in the evenings. The computer modeling would take some time, but it was another adventure, and he was eager to get started.

It was a shame to leave his friend Jason, but it was necessary. He had already had some guilt feelings about the vast changes Jason was making in his own life. He knew that each person was responsible for their own life yet still, he couldn't help feeling that without his influence maybe Jason would have stayed married and actually come to love his wife over time, but… No! he chided himself. Don't let yourself go down that 'what if' road.

Jason did whatever he had to do, just as each person did. Jason was his friend and even though he knew time and space were increasing between them with each paddle stroke, he also knew that friends' paths cross many times in a lifetime. He would see Jason again.

With good weather he could make the trip to New Orleans in twenty days.

Chapter 16

Jason arrived at the college a half hour early for the ACH tests. He was both apprehensive and excited at the prospect of being a student again. When he'd quit high school, he'd been glad to be done with it. At that time in his life, he'd had his fill of schoolwork. He'd lost interest in his lessons and became complacent. Plotting points on an X-Y plane or diagramming sentence structure seemed the most useless things in the world.

The factory had changed all that. He now realized that there were worse things in the world than the ABC's. He really felt like he'd just returned from Hell. He'd taken two classes since he's left high school and that was just enough to give him the minimum requirements for his high school diploma. Since that time, he'd wondered if he'd become more smart, dumber, or stayed about the same. Of course, he had just finished some lessons in time travel, but he doubted that he could use the time in the bar towards a college degree.

The director of the ACH test walked into the room five minutes before eight and passed out a booklet to each of the students. There were only a dozen students in the vast lecture room and after a brief explanation she announced, "You may begin," she then left the room.

Jason looked around, wondering if there were hidden cameras or a one-way window somewhere. From his high school days, he remembered tests, as long hours of concentration while the teacher paced up and down the aisle looking for cheaters and sneaks. He opened the booklet and immersed himself in their multi-

module tests. Two hours later the instructor reappeared and announced a fifteen-minute break. Jason went outside and squinted from the brightness of the fall colors. He was exhausted. Two hours down and two to go. He quietly passed gas thinking how strange it was to go two hours without farting. He knew he was doing fairly well as he'd gotten pretty deep into each module before running out of answers and rushing into the next section. He wished more than once that he'd paid more attention so many years ago. He wished he'd done the extra work that it took to get A's rather than C's but wishing wouldn't matter now. Others had done it; he could do it. Jason walked back to the lecture hall. The instructor said, "At twelve noon, please place the tests on my desk. Your results will be forwarded to your counselors. Modification in your schedule must be made within the first three weeks of school." She left the room.

Jason again buried himself within the modules and leaned hard to the task. At noon he placed his result on the desk with the others and went home.

His mother was setting in the kitchen when he returned home. She arose and kissed him on the cheek as always and cleared off a place at the table. "How about some ground meat and chips?" She said, as she busied herself with the kitchen chores.

Jason was famished. The ground meat was a homemade concoction from a Sunday pot roast. It was ground in a hand meat grinder and mixed with a boiled egg, sweet pickles, and some sort of homemade dressing. The only place he had ever eaten it was at home. The quick lunch was hungrily devoured. His mother entertained him continuously with the neighborhood news of deaths and illness and new births. All the names seemed vaguely familiar, but Jason was much better at listening than remembering. The next day she could re-tell it all and he'd probably not remember any of it, still, it was home. It was familiar and it was comforting to hear all the neighborhood tales.

He hadn't told anyone of his visit to the college or the array of tests he had been taking. He was afraid that the telling might jinx it, or that he might fail and then the embarrassment of failure would be too much for him. Now after the stress of the morning and the delicious of the lunch he felt ready to burst with news.

"Well son, what is it?"

"I just finished taking some placement tests at the college. I have a counselor that is placing me in some beginning classes, and I start class on Monday."

He waited for her reply, but the look on her face told him he was moving in the right direction, that she felt he should take, she was always there for support and acceptance but meddling in her children's affairs was beyond her. She was from a generation of women that believed in faith and prayer rather than in domineering interference. It was obvious from the look on her face that Jason had made her the happiest mother in the neighborhood.

It was a contagious feeling. It dispelled many of his fears. Somehow it helped lift a heavy load from his mind.

"When your father died, he left me a little money, and his life insurance was added to that. If you need any money you are welcome to it."

"My counselor thinks I might be able to qualify for a Basic Educational Opportunity Grant. If the BEOG comes through I should be able to scrape by. I'm behind in my child support and since I'll be quitting my job, I'll continue to get future behind. I don't know what will happen legally but I'm going back to school. I'm going back and I'm not quitting until I've finished."

She just smiled. This was great. This was bragging rights in the neighborhood. She wouldn't mention anything about the child support, her son was now a college student. How proud his father would've been.

Jason arrived at the college campus at eight A.M. He was an hour early and he wanted to get familiar with the building. There was a light fall drizzle that chilled his bones but not his spirits. He walked into the main superstructure and bought a vending machine cup of coffee. The heat flowed from the cup into his hands and the black hot fluid scorched his lips as he took a precautionary sip. He sat at a padded bench and leaned against the cement block wall, letting the new surroundings saturate his being. He thought about the good coffee he'd received at the old plexiglass stores window in the rolling mill. It already seemed a universe away. He'd thought about quitting the factory many times. How he'd walk up to all the supervisors he'd had problems with and give each one a good chewing out. How he'd bring up every stupid injustice and every idiotic remark and how he'd tell them all to kiss his ass. He'd even fantasized about slugging one of them.

How funny it all seemed now. The minute he'd decided not to go back, all the old grievances disappeared. The old problems didn't matter anymore. He'd heard that just before a man commits suicide a peacefulness comes over him and all his worldly problems dissipate. Jason felt he'd somehow cheated death. He had a peacefulness about him, and he was more alive than he'd ever been.

Jason looked down the aisle, out the glass doors and noticed a man on crutches approaching. The heavy-set, middle-aged man fumbled at the door and got it opened. He wedged a crutch into the base of the door and made his way across the threshold. Jason leaned forward as he felt the urge to help him, but before he was off the bench the guy had cleared the doors and was making his way down the aisle. He pulled his feet back to make sure he was out of the way when he saw the left crutch set down on a spot of water.

The crutch slipped on the slick tile and the man went down hard. His right leg separated at the knee and went spinning across the aisle. It was at first a gruesome sight. One of those things that happen, and your mind knows it can't happen, but it does. The Prospectus came to rest against the opposite wall. Jason hid his embarrassment, got up retrieved the plastic leg and returned it to its owner.

"Need this?" Jason asked as he handed the leg back to its owner.

"You two-legged people piss me off," Dick said in obvious frustration and anger, as he busied himself strapping his leg back into position.

Jason put his foot up against the main's good foot, braced himself and reached for his hand. He pulled him into an upright position and introduced himself.

"Jason Strider," he said.

"Dick Snyder's my name," the crippled man said, "If the school administration would keep these floors dry, things like this wouldn't happen."

He went on to explain what a crummy system it was. Jason stood and listened, realizing that what Dick was really complaining about was his own clumsiness, but was bound to blame everyone else as he vented steam.

Dick smoothed back his thick brown curly hair and brushed the dust off his pants and coat. "Buy you a coffee?" Jason asked.

Dick gave him a cautious look. A strange offer for a college student. "Sure," he answered, as they walked and hobbled to the vending machines.

"You must be new here," Dick said.

"First day, I'm supposed to meet my counselor here at nine AM. I'm just trying to get familiar with the place. You been here long?"

"Sometimes it seems like I've been here forever. I started six years ago."

"Six years? What are you studying?"

"History and languages. Mostly history, at least, that's what I enjoy. The languages are what keep me on my toes. Of course, keeping on my toes is easier for me. I only have half as many."

Jason let out a nervous laugh. He was always uncomfortable dealing with other's handicaps. The man, what was his name? Dick Snyder was sitting across the table slowly taking a sip of coffee. He was heavyset, with a few streaks of grey running through his wavy brown hair.

His face was slightly pockmark and his nose took up too much of his face. He had smile lines around his eyes from years of good times and his face was tempered with the pain of lessons long ago learned. Jason felt the same warmth he had felt when he'd first met LeRoi. They sat looking at each other for a few seconds. It wasn't uncomfortable. He wondered if Dick was feeling anything.

"Well," Jason said getting up, "I've got to be getting up to my counselors office. He's setting my schedule. See you around." Jason started to walk away.

"Hey Youngblood," Dick said, "Thanks for the coffee." Jason waved him off and continued, but the words echoed in his ears from a long ago dream and the memories threatened to overwhelm him. Get a grip he cautioned himself. Don't let your daydreams interfere with the real world. Don't get sidetracked. "I have only one goal," he said aloud to himself. "I will get a degree in engineering in four years, and the world must kill me to stop me. I will not be denied." He took a deep breath and entered WQ's office.

WQ Worthy sat quietly behind his desk. He was busily going over Jason's test scores. He looked up and invited Jason to have a seat.

"Not too bad," he said as he turned the pages. "You have a better than average math aptitude and your English is about average. You've tested high enough to begin classes. They will all count towards your degree in industrial engineering. I've placed you in Math 151A, English 113, Human Relations, and Drafting. The drafting is an optional course, but I figured you could use it to pick up some part time work at the Stout plant. We usually don't Co-Op freshman but with your industrial experience I figured you could do them a pretty good job. Your first math class begins in about," WQ looked at his watch, "twenty minutes." He pushed a large manila folder across the desk to Jason.

"Your schedule and everything you need to get started is in this folder. The math teacher will have a book you can borrow for the first week. Fill out all the paperwork and return it to me by Wednesday."

"To get to your first class, walk past the snack-bar area, turn left at the first aisle and go to room 127. It will be the third door on your left." WQ reached across the desk. "Good luck," he said as he shook Jason's hand. Jason didn't know what to say. He gathered up his folder, said thanks and headed for his first, very first class. Everything was happening so fast that he didn't have time to worry.

Jason walked back to the snack area and sat down across from his new friend Dick. He laid the manilla folder on the table and let out a long breath.

Dick sat and watched. He could see the nervousness, the doubts, the worry of the unknown. "relax if you can," he said in a joking manner. "You know Jason, sometimes I wish I could get back to where you are now. Back to where every minute was a new adventure. No, I can't get back there but it's almost the same just watching you sweat."

"Thanks a lot," said Jason, "If you want, I'll trade you. You go to my math class and I'll sit here and drink coffee." Dick just laughed.

"No thanks, I've got a history class in about fifteen minutes. Meet me back here after class though and I'll spring for the next round of black poison. Some of the other guys will be here and maybe we can put together a Euchre game." The small talk and Dicks relaxed mood was contagious and Jason relaxed a little. He glanced through his manilla envelope but quickly buttoned it up and excused himself and headed for his first class.

The room was clean and fresh, and he was the first one there. He picked a seat in the middle of the room and again shuffled through his folder. There was a pencil and a yellow legal pad. He pulled the pad from the folder, dated the top right-hand corner and started the first line with a question for himself. "Where are you now Jason Strider? What in the world have you gotten yourself into this time?"

The first ten minutes involved a brief orientation of the rules and grading procedure. The next thirty-five minutes reviewed everything he'd covered in his last Algebra and Geometry classes so many years ago. The last five minutes loaded him with work for the next class and then it was over. He walked back to the snack bar and located an empty table. He now had one borrowed book, one large manilla folder and one new future.

A new timeline had begun, and he knew he wouldn't have much time to worry about the past, the copper factory or the failed marriage.

One day at a time was his new motto. He wanted to look ahead to the future, but it was too complex. Now he couldn't even predict what the next hour would bring, let alone the next four years. The one thing he was sure of, was a fading image of a broken old man in a copper rolling mill, sitting at a wooden bench in a locker room with his bag cascading over the edge of the bench from a grey speckled bush of kinky hair.

He looked down at his hands to make sure they weren't covered with the dark freckles of age and he reached up and felt his full head of hair. It was as if he had to keep reassuring himself of where he was. Dick was nowhere in sight and Jason knew he'd have to meet him another day. He had some unfinished business to take care of at the copper mill.

He hadn't called in to work that morning and he knew the crew would all be there going through their daily routine. He suspected there would be some kind of disciplinary action waiting for him when he got there.

He suddenly felt the weight of a large broad sword hanging at his waist. He reached for it but there was nothing tangible there. But within the intangible was a strange feeling of heft and power. He reached again and felt the presence of his father, but his hand came away empty. Why a sword he thought. He didn't know, but it somehow comforted him. He left the college and drove to the copper mill.

He pulled the old '65 green, mid-engine GMC up alongside the gate house and walked in past the big, black guard. They exchanged greeting as usual, but this time Jason realized that each of the things he was about to do would not be done again. This was his last hello and goodbye. He somehow wanted to stretch them out and in fact they were being stretched out as time seemed to be slowing.

To the people he was to say goodbye to, time was normal. To Jason, time had slowed to a crawl. He walked to the personnel manager's door. It stood slightly ajar. The name on the door said William Comfer, Personnel Manager. Jason pushed it open.

"Come in Patch, I've been expecting you. Ken told me he was sending you down here. Have a seat and I'll call your union steward."

Jason thought about his union steward. He was one of the same guys that had fooled him into kicking an empty relay box so long ago. "Ken didn't send me down here, and I don't want to see my union steward."

"The contract gives you the right to have a union official present during any disciplinary meeting, and that's exactly what this is going to be."

"I'll give you some good advice Patch, you've got one warning coming for horseplay, and since you're obviously just getting here, you'll be getting number two for coming in late. Three strikes and you're out boy. One more warning and even a union steward wont' be able to help you."

"This is the most serious infraction in your last seven years and if I understand it correctly, someone could have been killed in your fiasco last Friday. Now do you still want to continue without the union steward?"

Jason sat and patiently waited for him to finish. The Personnel Manager was obviously talking about 'Jason the factory worker' and that person no longer existed. It was so easy to be calm.

On his old timeline from a week ago, Jason would have been arguing and alibiing and waving his arms and getting red in the face, but that person no longer existed. A new person sat in the old chair. A new man listened to facts that no longer mattered. Had they ever mattered?

Was all the worry and stress and sweat and misery real or was it all an illusion? If the copper mill and the hard labor was an illusion, then what was real? He drug himself back to the meeting. He had sat in this chair seven years earlier. He had been a green teenager just out of high school. He'd been eager to get into the workplace and show what he was made of, to right wrongs and accomplish good deeds. What had he done? Hard used was what he felt like. Hard was how he felt. Hardened in mind and body, but also strong.

"I came here to quit, Bill. If you want to write me up, go ahead and do the paperwork. I'll sign it."

Mr. William Comfer got a confused look on his face. He was accustomed and prepared for all the different reactions. He prided himself in controlling the situation no matter what the circumstances. Now he wasn't sure which path to take. Did the company really want to get rid of this rebellious young worker? Would this look good for him as a personnel manager, or would it somehow make him look bad? How could he use this to his own advantage? He wanted time to formulate a plan, time to work every situation to his best advantage.

Bill folded the paperwork and slipped it back into his top drawer. "Let's not make any hasty decisions here Jason. You've developed a lot of skills while you've been here. You're qualified on most every machine in the plant. You've become a pretty good sticker and Ken speaks highly of you."

"Why don't we forget about the warning for a couple of days? You go in and get started on today's run and we'll review this next Friday when everyone's had a chance to cool down. Maybe I didn't get the full story of how things went down last week. You know, I've never been one to rush into anything without looking at all the facts."

Jason couldn't believe his ears. Ten seconds ago, he was the worst worker in the factory and now all of a sudden, he was a highly skilled worker with some potential. He almost lost his composure. He wanted to shout hypocrite, but he wanted more to just be done with it. This guy was just another politician. Just another bag of worthless hot air.

"It's too late Bill. I'm already gone. I signed up for college and I'm going to become and engineer. Put together any papers you want me to sign and I'll stop and sign them on my way out. I'm going in to clean out my locker and say goodbye to a couple of the guys." Jason got up and turned to leave. A strange movement caught Bill's eye. It was the way Jason placed his hand on his waist when he turned to go, like he was putting something away.

Jason closed the door and walked toward the plant. He walked by the old-time clock, green and chipped from years of use. He walked along the shiny reels of copper wire.

"Hey Patch? You get lucky this morning?" Someone shouted from the multitude of noisy machines. Jason just kept on walking.

Bill had called ahead and got hold of Ken. The mill was humming along like a well-oiled watch. Ken couldn't believe the words coming from the personnel office.

"Quit? Jesus H. Christ! I fucking told you to write him up, not get rid of him. What the fuck am I going to do now? Stickers like him don't just come along every fucking day."

"Nothing I could do Ken. He had his mind made up when he walked in. I tried to stall him off, but it didn't work. Do you want me to send security up to walk him out?"

"Security? Why security, has he got a fucking gun or something? No, no fucking security. College? College? Jesus H. Christ." Ken slammed down the phone and ran his fingers through what was left of his thinning black hair.

He walked to the window and saw Jason walking toward the Hill with a coffee in his hand. Jason walked up to number one and sat the coffee on the dead man.

Jason felt out of place in his street clothes. He was out of place. The Hill was already slipping away. He felt drawn to it. Drawn to all the hands. Jason watched as a long red snake of hot copper hit the entrance guide. John reached down, caught it by the nose and passed it to T-Bone, who in turn passed it to the Hillbilly.

Jason stuck out his hand. "This be it for me big man. I'm done. I just came from personnel. Bill's making out the paperwork. Just wanted to say goodbye. Goodbye and thanks."

John stuck out his hand and felt the strength. He wanted to hold it. To hold Jason to the mill for a bit longer. The next bar hit with a thud and John spun and made the catch. He turned back to Jason. John saw the resolve in Jason's eyes. "You'd been a good one, if you'd just stuck with it," John said as he reached for another bar.

"Yeah, all I needed was a good teacher." They both smiled and Jason turned and walked away. He waved to T-Bone as he left the Hill.

Ken watched as Jason walked toward the locker room. Ken hurried down the steps. "Hey Youngblood," he yelled. Jason stopped and waited for ken to get to him.

"College huh?"

"I started today. In fact, right now I'm between classes. Just came back to say bye to the big man."

"You get some free time, come back and visit us." Ken stuck out his hand and Jason shook it. "Wished I'd had the balls to quit and go back to school twenty years ago," Ken said. "Good luck Patch."

Jason's locker waited for him. He opened the door for the last time. His tools leaned in the corner. An old holster, a belt, some tongs and side-cutters, a steel-handled hand ax, and one worn out pair of greasy work boots.

Nothing there to help him in his new life. Nothing to take with him. It was over. He shut the locker and walked away.

Time for Jason was now like a big ball rolling down a steep hill. The more momentum it collected the harder it was to slow down. Days filled with classes and projects, evenings filled with homework and study. He'd pick up drafting projects at the Stout plant and work on them at the college.

Upon completion of a project, he'd fill out his time and the college would pay him. The college would, in-turn bill the Stout plant, and everyone would make out. Jason was making valuable work contacts. WQ helped him over and difficulties that were outside the normal realm of his class material.

Jason had always dreamed of working at the Stout plant. The benefits were twice what the rolling mill paid and even though he did not yet qualify for the complete UAW package, he was certainly knocking on the door.

He had met many new people at school and though none were as close as his old friends, every day was an adventure.

Dick Snyder was a never-ending wealth of information. It seemed there was nothing he didn't know. His depth in history had led him to some interesting philosophies and Jason soaked all the various ideas and theories up like a sponge.

They were sitting at one of the snack tables in the cafeteria. They had finished a game of Euchre and Sandy moved over to their table. She was a tall brunette with soft features and a silky-smooth voice. She talked in a quiet sophisticated rhythm that always kept Jason off balance.

When she pulled up her chair, her lower leg brushed against Jason's. Neither pulled away but she kept up a friendly banter as if she weren't noticing the warmth that flooded through Jason. It spread up his leg and set his groin on fire. It was a slow burn that radiated inward.

She rested her hand on his forearm and talked of easy classes, harsh professors and pretty clothes. It was all words flying off into space for Jason. He responded as if on automatic pilot. He had no idea what he was saying but it must somehow be making sense. No one seemed to notice the flashes running through his loins and exploding somewhere in his brain. He looked across the table at Dick, but Dick just sat back with a quiet smile and said nothing.

"My ride seems to have stood me up Jason. Do you think you could give me a ride home? I'm done with my classes and I don't feel like riding the bus."

Jason couldn't believe his ears. He had been frantically searching for a way to get her into his van, but his brain wouldn't kick out a solution. The question hung there waiting to be answered. No one else seemed to be noticing. Jason put his books and his jacket on his lap and stood up, hoping no one would notice. Hoping he could somehow relax enough to stand in an upright position. "Let's go," he said. She prattled along as they walked to the parking lot. He opened the van door for her, and she entered. He got in on his side and started it. He drove across the bridge and pulled to the edge of the river across from the campus. The heater was on high and Jason reached behind the seat to a storage compartment, pulled out a blanket and threw it in the back. He locked his door, reached across her and locked her door and reached around behind her, with his hand caressing her neck, he lightly kissed her. She held him for a moment and then relaxed in his arms. They sat there for a while, feeling the closeness of each other. They felt the warmth of the heater flood through the old metal van and the chill of the winter day slipped away.

Jason was sitting on the engine compartment between the two front seats. Jason pulled his knees up to his chest and did a reverse roll into the back of the old van. The rear springs recoiled from the shock of his body hitting the thin carpet which covered the rear floor of the van. He came to rest in a cross-legged sitting position and smiled up at her.

"Why Jason Strider, what sort of a girl do you think I am?" she said, as she laughed out the words, faking a slow southern drawl

He reached up beneath her blouse, feeling her firm breasts and roughly dragged her back with him. They fell together laughing and playing in a wrestling match that soon slowed to deep passion.

Deep auburn hair hanging in his face, the firmness of her body, the fire of his depth and the need of both of them exploding in rough rasps and gentle sighs of ecstatic pleasure…and then the quiet warmth of two bodies that had for a brief few seconds become one. Of the union of the ages. Of the strength of two separate universes briefly and completely yielding to all that God had made of them.

As the real world slowly crept in around them, they quietly put themselves back together. He climbed back to the driver's seat and let out a long sigh. How long had it been? What an unexpected pleasure.

She climbed up over the engine compartment. "Where's home?" he asked.

"Just take me back to the parking lot. That's where my car is." Jason nodded and put the old van in reverse. It didn't bother him in the least that she hadn't really needed a ride…or had she. He smiled a deep knowing smile.

Sometimes heaven sits at a table next to you and you don't even know it. She directed him to her car and then jumped down out of the van. "See you tomorrow," she said, and then she was gone. Jason had another class to go to, but the math facts didn't seem to make much sense.

Dick sat at the snack area table that had become a permanent Euchre table for a number of students. It was a good place to sit and discover new truths. Jason walked to the table and sat down, dropping his books and school paraphernalia to the floor. His carrying bag which held his array of schoolbooks was an old WWII gas mask side pack. It had belonged to his uncle and had somehow ended up among his father's possessions. It was big enough for two or three books and a few tablets. The old green canvas bag was tattered and worn, but functional and Jason felt a comfort with it that he couldn't find in the typical bookbags and leather briefcases that most of the students carried.

Dick sat at his table, slowly shuffling the faded deck of cards that continued to provide cheap entertainment for the various players that always seemed to materialize within a few minutes.

Jason pulled out a lottery ticket and flashed it in front of Dick's eyes. "This ticket is going to win for me the tidy sum of one point nine million dollars."

Dick hardly took notice of the ticket. He looked at Jason and shook his head. "Why do you bother with such nonsense? Don't you know that you already won? Aren't your professors teaching you anything?"

Jason put the ticket back in his pocket. "What do you mean I've already won? I sure don't feel like I've won anything, in fact, I was barely able to come up with this semesters tuition."

"The reason you don't think you've won is because you don't know anything about money. If you did, you wouldn't bother to play the lottery game."

Jason was curious so he decided to play along. He knew he was broke, and no amount of talking would put a dime in his pockets. No amount of philosophy would make a hungry man full and no matter what Dick was about to say, he knew that he had never won a lottery.

"You're a lucky man, Jason Strider, you've beaten the odds many a time and you hardly ever notice it. What were your odds of getting out of that copper factory after seven years?" Dick continued without waiting for an answer. They weren't good, yet here you sit, big as life. But the lottery you won was much longer odds than you getting out of the copper mill. The lottery you won carried odds so enormous that all else pales in comparison." Dick took a deep breath and rearranged his artificial leg. He scooted himself back up in his chair. He seemed always to be slipping into a slouching position. He absent-mindedly continued to shuffle the cards and finally began to speak. "You are old enough that you should begin to understand about money. Seldom do people understand the elusiveness they seek."

Jason sat back and listened to what he would forever remember as 'Dicks Lottery Story'.

Dick began, "let's begin at a time before there was any such thing as money, and let's pretend you are a farmer, and you own a couple of cows. The farmer down the road owns a couple of hogs. You want one of his hogs, and he wants one of your cows. The problem is, you know, and he knows that one of your cows is worth more than one of his hogs. You agree to make a deal, but since your cow is worth more, you accept the hog plus an IOU for half a hog, as he doesn't want to give up both his hogs." Dick paused a few seconds and then asked his question. "What is money in this situation?"

Jason didn't need time to think. "The IOU is the same as money." He answered.

Dick continued, "What does the money represent?"

"This is pretty elementary stuff Dick; the money represents half a hog."

"Well Jason that's a good answer but it's wrong. That's the same answer everybody always gives, but it's just plain totally wrong."

Jason ran the simple scenario through his mind, but the question was too simple to get a wrong answer. He had been fooled and tricked enough that it didn't bother him. "If that's the wrong answer, then you probably haven't given me all the clues."

"The clues are all there, the problem is everyone gets confused by thinking the IOU is equal to half a hog. The IOU is really only equal to the time the farmer has spent on half a hog."

"We find material things in which to invest our time and then we trade those material things for other material things, and we balance out the difference with something we call money. After a while people get confused with whatever they are trading and lose sight of the fact that what they are really, truly trading is their time, for someone else's time."

Dick sat back in his chair and waited. He was acting as if he had just revealed some great truth. Jason wasn't buying it. It was too simplistic, and besides, it always pissed him off when someone said he answered a simple question wrong.

"What about gold?"

"It represents the time it took someone to file a claim and dig it out of the ground."

"What about a house?"

"It's worth the time it took someone to build it."

"My van?"

"The time it took to produce it."

"My schoolbooks?"

"The time it took to write them, the time it took to make the paper, and the time to make the ink. Nothing tangible in this world is worth any more than the time that someone else has in it. People kill over money but what they are really doing is killing for someone else's time."

"The reason people get so upset about losing possessions is that possessions, including money, only represent the time you have invested in them. All manufactured goods, all money represents other people's past hours.

"If people realized what they were stealing, they might not be in such a hurry to steal it. On the other hand, it's easy to see why people try so hard to hang on to their possessions. Their possessions, including their money represents their past. You worked seven years in a copper factory. Did you save any of your past?"

Jason just laughed, "I tried to hang on to some of my past, but my ex-wife and her lawyer got it all. They'd like to get a piece of my present and my future too, but I quit dealing with them. I no longer respond, and I refuse to cooperate. If what you are saying is true, then those people are time bandits."

"That's correct. You probably also noticed that when someone relieves you of all your possessions it really frees you or lightens your load so you can more easily get on with your present and your future. Time is the only real commodity in this world, and it is the only thing of any value. Possessions only represent other people's time or a piece of your past. If a person is creative and spends his time wisely people will save his time and trade it again and again. A true craftsman's time may be traded and become more valuable as time goes on.

"Because money is a note that represents a time debt, the more money you give for something the more you are saying that a particular person from some time in the past has created something of some value."

"Thieves, or time bandits are simply people with such low

self-esteem or such limited skills that they risk a big part of their future for someone else's past. In other words, they risk going to jail, or they risk being labeled a thief trying to acquire a bit of useless material, when if they looked at it logically, they could probably figure out how to create what they are stealing with a small investment of their own time."

Jason had been sitting there listening when he noticed the lottery tickets sticking out of his pocket and remembered that that's what had created this rather lengthy recitation. "So, what's all that got to do with me winning the lottery?"

"Well, you need to know what is real and what is illusion. If people don't understand what money is, they may chase it all their life and then be disappointed because they never attained any of it, or even more disappointed when they attain it and realize there is no happiness with the accumulation. I heard a preacher say one time, 'If you can't be happy today, then you can't be happy.' I usually don't quote preachers, but this particular one went on to tell the old story about a young couple walking down a boardwalk next to the ocean."

Dick continued with the story. "She was a young beautiful girl of seventeen and he was a mere eighteen years old. They walked hand in hand and their love for one another was obvious from the way they walked and talked. It was one of those scenes that when you look at them you have to smile to yourself because most everyone has been there or dreamed of being there. Anyway, as they walked along the beach, absorbing the sounds of the ocean and drinking in the warm sun on their young bodies, they passed by a beautiful mansion. It was one of the old eighteenth century creations, laid up from native stones of granite and decorated with gingerbread woodworking's, and magnificent wood columns of oak and walnut. The young girl looked up at the mansion and squeezed her lover's hand. 'If I could but live in that grand house, I would be the happiest woman in the world.'"

"Up in the house peering out from behind a fancy white curtain of lace sat an old, widowed lady. Her hands were wrinkled and weak, her husband long ago dead. She watched through thick glasses, the young couple walking past her mansion. Her vision was almost gone but through the thick lenses of her wire rimmed glasses, she watched the young lovers pass."

"'If I could change places with that beautiful young girl, I would be the happiest person in the world," she said to the empty room.

Dick paused and thought about the story. Jason had sat there and listened. "When you came in with your lottery ticket, you reminded me of that story. I first heard it from the old preacher at a church I used to attend, but it all kind of fits into a scenario I've been thinking about."

"When you buy a lottery ticket, you invest a little of your time with the hopes of winning all those little bits of time that everyone else has invested. Even if you win, you will only be winning a million little pieces of everyone else's past. All that past baggage would probably just weigh you down and keep you from completing whatever it is you're trying to complete."

Jason continued to listen, but it was obvious he was not convinced of anything. "You mean to say that if you won this lottery, you wouldn't accept the money?"

Dick shook his head, "No, I didn't say that. I'm just trying to lay the groundwork so you will understand what is valuable and what is not so valuable. The lottery dollars only represent the past. What's more important to you, other people's pasts or your future?"

"My own future is the most important, but I'm only saying that because I want you to get to the part that explains how I've already won the lottery. I do have another class to go to today."

Dick rearranged himself again and took a sip of coffee. He doodled with his pen as he spoke, making little spirals and loops in his notebook.

"How old are you Jason?"

"Twenty-seven."

"How old was your mother when you were born?"

Jason did a quick calculation in his head. He was the firth of seven children. His mother was thirty-one when he was born.

"Thirty-one," he said, now becoming more interested in the conversation as it was being directed more at himself.

"Do you know how old your mother was when she created the egg that created you?"

Jason squirmed a little bit. He didn't know much about the technicalities of fertilization. Sexuality in general wasn't a topic he ever spoke of, especially when it might concern his parents. He also didn't see any connection between his lottery ticket and the direction this conversation was going. He looked around uncomfortably, checking to see if any of the nearby students were listening. No one seemed to be interested.

"No," he replied, "How would I know how old my mother was when I was created? Probably thirty-one minus nine months."

"Good answer Jason, but you're way off. The eggs that a woman carries are all created when she herself is just a fetus. That means you are twenty-seven years, plus the nine months before your birth, plus the thirty-one years that your mother was when you were born, plus the six months previous to her birth." Dick did a quick calculation with his pen and continued. "That means you are now somewhere around fifty-nine years old." Dick paused, waiting for the information to sink in, and then continued. "At that time, way back fifty-nine years ago, your mother produced over six million eggs. Twelve or thirteen years later, her body began to release one egg a month. By the time she was thirty-one years old, her body had released about," Dick did some more quick figuring, "two hundred and thirty eggs from the original six million. As you mentioned, you have three older brothers and two older sisters."

Jason was becoming more interested all the time, as the mathematics, mixed with the timing was beginning to lay an interesting mosaic of life that he had never considered before.

"So, I have been chosen from one of six million eggs. How many sperm were involved?" He unconsciously squirmed again at the thought of speaking so clinically about something so personally discrete and private as his parent's sexuality. He was also having trouble accepting his new age of almost sixty. It reminded him of the old grey-haired man on the bench, back at the copper factory.

"Well, there are about one hundred and twenty million sperm involved in each ejaculation. The sperm are not so old as the eggs. They are created by the testicles about one month before they are used."

Jason blushed slightly at the thought of his father's sperm. How often had he joked about ejaculate? He had referred to it as cum, load, wad, and a multitude of other terms that now seemed somehow obscene. He had seen sperm under a microscope once in a high school biology film. They had looked like little tadpoles with long tails. They were one hundred and twenty million strong, like soldiers going in search of the elusive egg. He thought of the many times he'd jerked off in his life. Boom, another one hundred and twenty million tadpoles down the toilet. Pow another hundred and twenty million sperm shot into space.

Dick continued, bringing Jason out of his thoughts and back to the present topic. "So, the population from which we shall construct our statistical data base contains (A), six million eggs, and (B), one hundred and twenty million sperm."

"Our experiment involves drawing one of these sperm and one of these eggs and combining the two. This new creation will be called Jason Strider. The event is both independent and unique because the sperm that created you was unique among the one hundred twenty million, and the egg was unique among the six million from which it came. If it had been any other egg or any other sperm, you would not be you, you would be someone different."

"The timing was also critical, but I will leave out the variables concerning time, as we already have quite enough of an improbable event to prove my point."

"So, what's your point?"

"My point is that the probability of (A), plus the probability of (B) occurring is one over six million plus one over one hundred twenty million. The chance that you would be born as you, were one in one-hundred twenty-six million. The odds of you winning a million dollars in the lottery are only one in three million."

"If you're alive, you've already won the lottery. You've won a chance to spend seventy to eighty years any way you please. You are only twenty-seven years old now, so you have another fifty or so years to spend your time, your winnings, on whatever you desire, and we both know that time is the only thing of any value."

Jason mulled it over in his mind before responding.

"That doesn't seem right. If the odds are that great against me being born, then why did my mother have seven children? It seems to me if it was that difficult to have a child, then there wouldn't be so many people in the world."

"The odds are good that your mother would have children, the millions of sperm combined with the multitude of eggs over the years makes it probable that some children will be born. But the odds that you would be you are not good at all."

"So, you're saying that since I'm alive, I have already won the lottery."

"That is correct. You not only won, but you have also won big and you have no debts to pay off. Your future belongs entirely to you. You're blessed with time. There will come a point in your life when you'll wish you had but a few more minutes. When that time comes, you'll realize that no amount of money or gold can give you that which you will ask for."

"OK, I see your point, but somehow I still don't feel any richer."

Dick just laughed, He rearranged his leg again and thought a moment about how little was left of his own timeline. The injury that had cost him his leg, had left him with a bone disease that was not only shorting his bad leg, but it was also slowly shorting his timeline. Nothing could be done, nothing but try to relay some of his thoughts to someone that might keep them alive or pass them on, for a person could continue to live if his thoughts were kept alive. Even though his body might go, his ideas, if they were good enough, could keep him alive. After all, wasn't Da Vinci alive through his sculpture? Wasn't van Gogh alive through his portraits? Yes, according to that type of logic, a man could live forever if his ideas were pure enough.

Lightning Source UK Ltd.
Milton Keynes UK
UKHW030657021122
411507UK00001B/33